Molecules and Cells

revised edition

John Adds • Erica Larkcom • Ruth Miller

Series Editor: Martin Furness-Smith

Endorsed by

First published in 2000 by:
Nelson Thornes Ltd
Delta Place
27 Bath Road
CHELTENHAM
GL53 7TH
United Kingdom

This edition published in 2003

05 06 07 / 10 9 8 7 6 5 4 3

A catalogue record for this book is available from the British Library

ISBN 0 7487 7484 X

Illustrations by Hardlines and Wearset Ltd
Page make-up by Hardlines and Wearset Ltd
Printed and bound in Croatia by Zrinski

Acknowledgements

Photographs

Biophoto Associates: 4.4;
Erica Larkcom 3.10a, b
John N. Adds: 4.2, 4.5a;
Oxford Scientific Films: Kent Wood 5.11;
Science Photolibrary: K R Porter cover, Andrew Syred 1.14, 4.18a, CNRI 1.15b, 5.7 bottom, Claude Nuridsany & Marie Perennou 1.17, Dr Gopal Murti 2.13, Secchi-Lecaque-Roussel-UCLAF, CNRI 4.6b, Dr Don Fawcett 4.11b, Bill Longcore 4.12a, Dr Jeremy Burgess 4.13a, 4.15, Dr Linda Stannard U C T 4.17a, E M Unit Southampton University 4.18b, Dr Robert Dourmashkin 4.25, 5.3, Biophoto Associates 5.4, Science Source 5.7 top, Professor G Schatten 5.8, J C Revy 5.9a, b, c, d, Rosenfeld Images Ltd 5.12.

Artwork and other material

Dean Madden, National Centre for Biotechnology Education, Reading University: Figure 3.9 (from NCBE newsletter); Figure 3.11 (original in EIBE, Unit 1); Table 3.2 (modified from NCBE poster).

Contents

4 Cellular organisation 51

5 The cell cycle 72

Introduction

This series has been written by Chief Examiners and others involved directly with the Edexcel Advanced Subsidiary (AS) and Advanced (A) GCE Biology and Biology (Human) specification and its assessment.

Molecules and Cells is one of four books in the Nelson Advanced Science (NAS) series. These books have been developed to match the requirements of the Edexcel specification, but they will also be useful for other Advanced Subsidiary (AS) and Advanced (A) courses.

Molecules and Cells covers Unit 1 of the Edexcel specification, the content of which is common to AS GCE and A GCE Biology and Biology (Human). Important basic concepts, fundamental to any Biology course, are introduced and a sound foundation is laid for the study of topics in other Units. The contents include:

- the structure and roles of some biologically important molecules
- the structure and roles of enzymes, together with the factors that influence their activity
- a basic understanding of the genetic code and protein synthesis
- the organisation of prokaryotic and eukaryotic cells
- the cell cycle and the process of mitosis.

The other student books in the series are:

- *Exchange and Transport, Energy and Ecosystems*, covering Units 2B, 2H and 3
- *Respiration and Coordination*, covering Unit 4 and including the Options
- *Genetics, Evolution and Biodiversity*, covering Units 5B and 5H.

Other resources in this series

NAS *Tools, Techniques and Assessment in Biology* is a course guide for students and teachers. For use alongside the four student texts, it offers ideas and support for practical work, fieldwork and statistics. Key Skills opportunities are identified throughout. This course guide also provides advice on the preparation for assessment tests (examinations).

NAS *Make the Grade in AS Biology with Human Biology* and *Make the Grade in A2 Biology with Human Biology* are Revision Guides for students and can be used in conjunction with the other books in this series. They help students to develop strategies for learning and revision, to check their knowledge and understanding and to practise the skills required for tackling assessment questions.

Features used in this book – notes to students

The NAS Biology student books are specifically written to help you understand and learn the information provided, and to help you to apply this information to your coursework.

The **text** offers complete and self-contained coverage of all the topics in each Unit. Key words are indicated in bold. The headings for sub-sections have been chosen to link with the wording of the specification wherever possible.

In the margins of the pages, you will find:

- **definition boxes** where key terms are defined. These reinforce and sometimes expand definitions of key terms used in the text.
- **questions** to test your understanding of the topics as you study them. Sometimes these questions take the topic a little further and stimulate you to think about how your knowledge can be applied.

Included in the text are boxes with:

- **background information** designed to provide material that could be helpful in improving your understanding of a topic. This material could provide a link between knowledge gained from GCSE and what you are required to know for AS and A GCE. It could be more details, information about a related topic or a reminder of material studied at a different level.
- **additional** or **extension** material which takes the topic further. This material is not strictly part of the Edexcel specification and you will not be examined on it, but it can help you to gain a deeper understanding, extending your knowledge of the topic.

INTRODUCTION

In the specification, reference is made to the ability to recognise and identify the general formulae and structure of biological molecules. You will see that we have included the structural chemical formulae of many compounds where we think that this is helpful in gaining an understanding of the composition of the molecules and appreciating how bond formation between monomers results in the formation of polymers. It should be understood that you will not be expected to memorise or reproduce these structural chemical formulae, but you should be able to recognise and reproduce the general formulae for all the molecules specified.

The chapters correspond to the sections of the specification. At the end of each chapter, you will find the **practical investigations** linked to the topic of the chapter. These practical investigations are part of the specification and you could be asked questions on them in the Unit tests. Each practical has an introduction, putting it into the context of the topic, and sufficient information about materials and procedure to enable you to carry out the investigation. In addition, there are suggestions as to how you should present your results and questions to help you with the discussion of your findings. In some cases, there are suggestions as to how you could extend the investigation so that it would be suitable as an individual study.

At the end of the book, there are **assessment questions**. These have been selected from past examination papers and chosen to give you as wide a range of different types of questions as possible. These should enable you to become familiar with the format of the Unit Tests and help you to develop the skills required in the examination. **Mark schemes** for these questions are provided so that you can check your answers and assess your understanding of each topic.

In this book there is an **Appendix**, which provides the **physical science background** that you need in the study of the Biology and Biology (Human) specifications. Some of this material is covered in the main text and this is indicated by a cross-reference. Much of the information in the appendix may have been covered at, or before, Key Stage 4, but it is a useful reminder of some basic scientific concepts.

Note to teachers on safety

When practical instructions have been given we have attempted to indicate hazardous substances and operations by using standard symbols and recommending appropriate precautions. Nevertheless teachers should be aware of their obligations under the Health and Safety at Work Act, Control of Substances Hazardous to Health (COSHH) Regulations, and the Management of Health and Safety at Work Regulations. In this respect they should follow the requirements of their employers at all times. In particular, they should consult their employer's risk assessments (usually model risk assessments in a standard safety publication) before carrying out any hazardous procedure or using hazardous substances or microorganisms.

In carrying out practical work, students should be encouraged to carry out their own risk assessments, that is, they should identify hazards and suitable ways of reducing the risks from them. However they must be checked by the teacher. Students must also know what to do in an emergency, such as a fire.

Teachers should be familiar and up to date with current advice on safety, which is available from professional bodies.

Acknowledgements

The authors would like to thank Sue Howarth for her help and support during the production of this book.

Examination questions are reproduced by permission of Edexcel.

About the authors

John Adds is Chief Examiner for AS and A GCE Biology and Biology (Human) for Edexcel and Head of Biology at Abbey College, London.

Erica Larkcom is Deputy Director of Science and Plants for Schools at Homerton College, Cambridge, and a former Subject Officer for A level Biology.

Ruth Miller is a former Chief Examiner for AS and A GCE Biology and Biology (Human) for Edexcel and former Head of Biology at Sir William Perkins's School, Chertsey.

Molecules

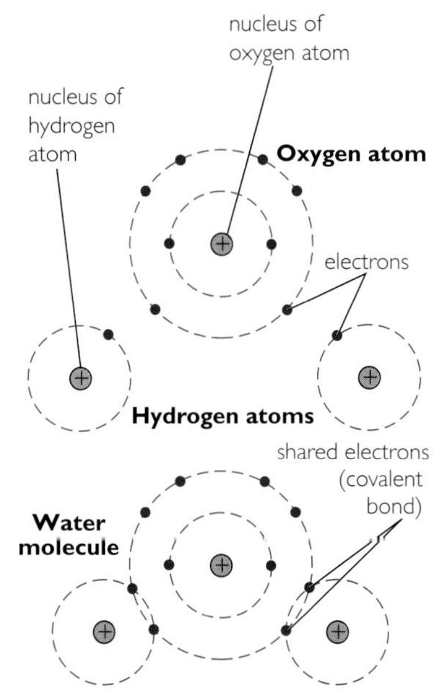

Figure 1.1 Atomic structures of oxygen and hydrogen, showing how they are combined in a water molecule.

Role of biological molecules

In order to gain an understanding of the nature of metabolic reactions within cells, it is helpful to know something about the structure and properties of the molecules involved. Most of the groups of molecules considered here are organic and all are of great significance in the structure and functioning of living organisms. (For a detailed explanation of the structure of atoms and molecules, see *Appendix: Physical science background*.)

Metabolism is the term used to describe all the reactions taking place within cells. These reactions can be divided into **anabolic**, in which compounds are being built up, or synthesised, and **catabolic**, where compounds are broken down. Anabolic reactions require energy whereas catabolic reactions often result in the release of energy.

Water

Water is necessary for life to exist on Earth. It is the medium in which all chemical reactions occur in living organisms. It makes up between 60 per cent and 95 per cent of the fresh mass of living organisms and is an important chemical constituent in all cells. Its dipolar nature and ability to form hydrogen bonds are responsible for its major role as a solvent and transport medium. In addition, it is also the habitat for a large number of organisms.

Structure of the water molecule

In order to appreciate the special properties of water molecules, it is helpful to know about their structure. In a water molecule, two **hydrogen** atoms are joined to an **oxygen** atom by covalent bonds. The oxygen atom has a nucleus, containing eight positively charged **protons** and eight **neutrons**, surrounded by eight negatively charged **electrons** (Figure 1.1). The hydrogen atom has one proton in its nucleus, plus one electron. In atoms, the electrons can be thought of as arranged in shells around the nucleus. The first, or inner, shell is the smallest and can hold up to two electrons, the second up to 8, the third up to 18 and the fourth up to 32. An atom is particularly stable when its outermost shell is full.

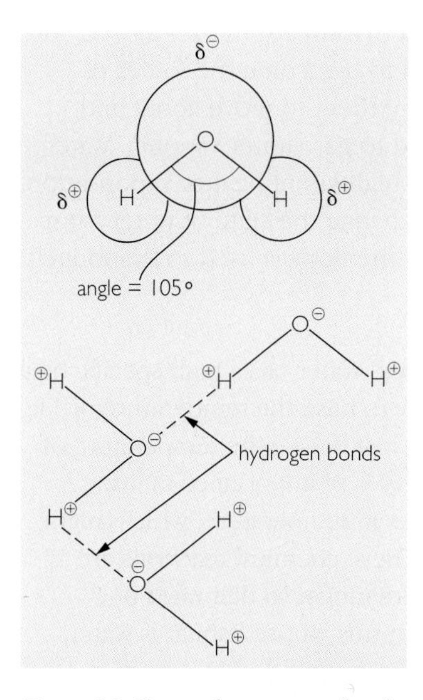

Figure 1.2 Shape of a water molecule and formation of hydrogen bonds between electrically charged hydrogen and oxygen atoms.

Atoms vary in size and do not all have four shells of electrons. The electrons in the oxygen atom are arranged in two shells: the inner shell has two and the second, outer shell has six. In each of the hydrogen atoms, there is a single electron in the first shell. The outer shell of the oxygen atom can hold up to eight electrons, so when the oxygen atom combines with the two hydrogen atoms, the electrons of the hydrogen atoms are *shared* with the oxygen atom, forming two **covalent bonds**. The water molecule formed is a stable molecule, but because of the arrangement of the two hydrogen atoms it has a triangular rather than a linear shape.

When covalent bonds are formed, the electrons are not always shared equally. In a water molecule, the oxygen nucleus attracts the electrons more than the

> **DEFINITION**
>
> A **hydrogen bond** is a force of attraction that forms between a very electro-negative atom (usually oxygen, but can be nitrogen or fluorine) and a hydrogen atom that is covalently bonded to another strongly electronegative atom.

hydrogen nuclei do. This results in the oxygen atom having a slight negative charge (δ^-), while the hydrogen atoms have slight positive charges (δ^+). This uneven charge distribution results in the molecule being **polar** and, because it is positive at one end and negative at the other, it is referred to as **dipolar**. Because water molecules are polar, they have an attraction for each other and form hydrogen bonds with neighbouring molecules (Figure 1.2).

Biologically important roles of water

As a solvent: Water is an efficient solvent for polar substances, such as salts, simple alcohols and sugars. When salts such as sodium chloride dissolve, the sodium ions and chloride ions separate and become surrounded by water molecules. The positively charged sodium ions **(cations)** are attracted to the negatively charged oxygen atoms of the water molecules and the negatively charged chloride ions **(anions)** are attracted to the positively charged hydrogen atoms (Figure 1.3).

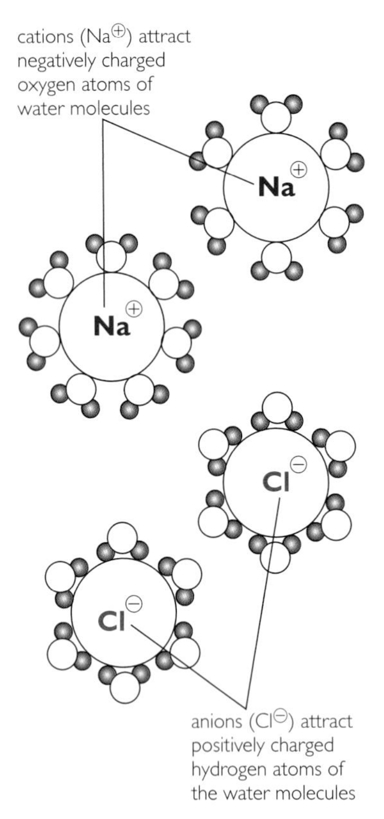

Figure 1.3 Distribution of water molecules around sodium and chloride ions in solution.

The organic molecules such as alcohols and sugars have polar **hydroxyl** (OH) groups to which water molecules are attracted in the same way. Non-polar molecules, such as lipids, and non-polar groups on other organic molecules do not dissolve in water.

The majority of metabolic reactions in cells take place in aqueous solution. The compounds involved are more chemically active in solution, since the molecules are able to move about more freely.

As a coolant: Because of the very strong attraction between water molecules, giving rise to the formation of hydrogen bonds, water has a higher freezing point and boiling point than other molecules of similar relative molecular mass. Methane (relative molecular mass 16), ammonia (rmm 17) and hydrogen sulphide (rmm 34) are all gases at 0 °C, but water (rmm 18) freezes at 0 °C and boils at 100 °C. A great deal of energy is needed to overcome the forces of attraction between the water molecules, allowing them to move about and change from solid (ice) to liquid and from liquid to gas (water vapour). Water can be a very effective coolant, because it has a high latent heat of vaporisation. This means that it takes a great deal of heat to change the state of water from liquid to gas. When we sweat, heat energy from the body is used to evaporate the water in the sweat, cooling us down.

To maintain relatively constant temperatures: Water has a high specific heat capacity, indicating that it requires a lot of energy to raise the temperature of 1 kg by 1 °C. Conversely, a lot of heat energy must be lost before the temperature of the same mass of water falls by 1 °C. This property is of importance in living organisms, because it means that sudden changes in temperature, which might upset metabolic reactions in cells, are avoided. These chemical reactions are allowed to take place within a narrow temperature range, so that rates of reaction are more constant. In addition, for organisms whose habitat is water, large fluctuations in the temperature of their environment do not occur.

For insulation: Water has its maximum density at 4 °C. As the temperature of a body of water, such as a pond, drops the colder water is at the surface and when it freezes, ice forms on the surface. The ice insulates the water below, enabling aquatic organisms to survive. This is especially important in cold climates and where there are cold seasons.

For transport: Water molecules are very cohesive, sticking together because of the presence of hydrogen bonds. This cohesive property is important in the transport of materials in solution within organisms. Examples are seen in the transport of ions in solution in the xylem of plants and in the transport of the soluble products of digestion in the blood plasma of animals.

Surface tension: At a water–air interface, the cohesive forces between the water molecules, the result of hydrogen bond formation, act in an inward direction, giving the water a high surface tension. This force causes the surface of the water to occupy the least possible area. High surface tension enables small organisms to land on the water surface and to move over it.

As a reagent: Water is an important reagent in metabolic reactions, particularly in hydrolysis, and as a source of hydrogen in photosynthesis.

(For more details on chemical bonds and the properties of water, see *Appendix: Physical science background*.)

Organic molecules

Organic molecules contain **carbon**, and in order to gain an understanding of the chemical nature of these molecules it is helpful to know a little of the structure and chemistry of the carbon atom.

The carbon atom is small with a low mass. It has six electrons arranged in two shells around the nucleus. There are two electrons in the inner shell and four in the second, outer shell. The nucleus in the centre contains six protons and six neutrons.

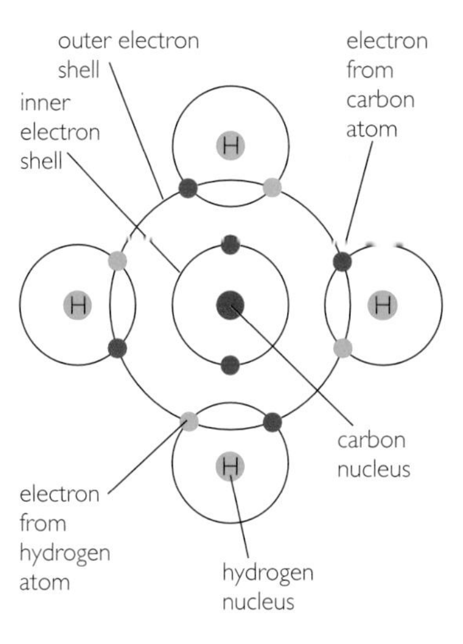

Figure 1.4 Structure of a molecule of methane gas, showing shared electrons in the orbits of the carbon and hydrogen atoms.

Because the outer shell of the carbon atom is not full, having only four electrons, it can acquire more by sharing electrons with other atoms and forming stable covalent bonds. A simple example of this is shown by the structure of methane, CH_4, where a carbon atom shares electrons with four hydrogen atoms (Figure 1.4). Each hydrogen atom has a single shell in which there is one electron, so when it combines with the carbon atom, its electron is shared with the outer shell of the carbon atom, forming a bond. In this way, the outer shell of the carbon atom acquires four shared electrons and is filled, forming a stable compound.

Covalent bonds are important in the formation and structure of organic molecules because they are strong and stable. When a carbon atom is joined to four other atoms, or groups of atoms, the four bonds are arranged spatially, forming a tetrahedral shape as shown in Figure 1.5.

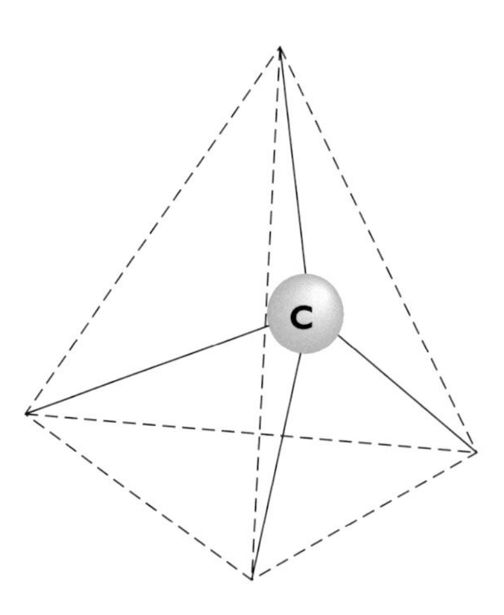

Figure 1.5 Tetrahedral arrangement of bonds around the central asymmetric carbon atom.

If you look at the formulae of most organic molecules, it is noticeable that they contain more than one carbon atom. The carbon atom is unusual in that it can bond to itself as well as to other atoms. This can result in molecules with straight chains, as seen in fatty acids, branched chains, as shown in the amino acid **alanine**, and in ring structures, such as α-glucose (Figure 1.6).

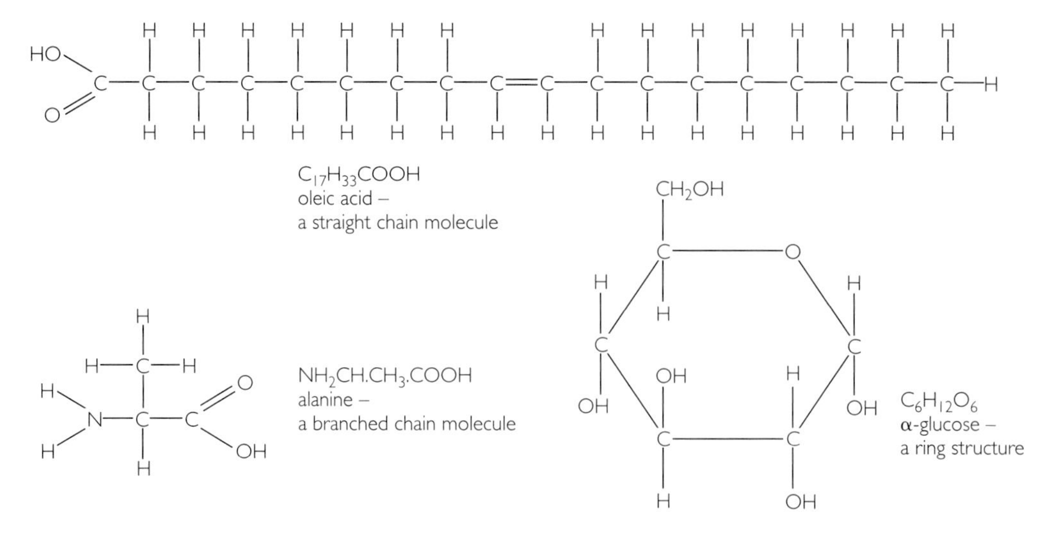

Figure 1.6 Structural formulae of oleic acid, alanine and α-glucose to illustrate straight chains, branched chains and ring structures. These structures occur because carbon can bond to itself as well as to other atoms.

Carbon atoms can also form **double** or **triple** bonds, in which two or three pairs of electrons are shared. These bonds can be formed with other carbon atoms (C=C and C≡C) or with oxygen and nitrogen (C=O and C=N). Triple bonds do not often occur. In Figure 1.6, the oleic acid molecule shows an example of a C=O bond and a C=C bond.

Carbohydrates

Carbohydrates form a large group of molecules that can be synthesised by plants. They contain carbon, together with hydrogen and oxygen. The ratio of hydrogen to oxygen atoms in the molecules is usually 2 : 1.

Carbohydrates are common constituents of plants, making up to 90 per cent of their dry mass. The cell walls are composed of **cellulose** (a polysaccharide), energy is stored in the form of **starch** (a polysaccharide), the products of photosynthesis are transported internally as **sucrose** (a disaccharide) and the energy source for metabolism in the cells is **glucose** (a monosaccharide). Animals require carbohydrates in their diet and obtain them, either directly or indirectly, from plant sources.

Carbohydrates can be divided initially into two major groups: the sugars and the non-sugars (polysaccharides). The sugars can be further separated into simple sugars, the monosaccharides, and the compound or double sugars, the disaccharides (Table 1.1).

Monosaccharides

Monosaccharides contain carbon, hydrogen and oxygen in the ratio 1 : 2 : 1, so their general formula becomes $(CH_2O)_n$, where n can be any number between 3 and 9. All monosaccharides also contain a C=O (carbonyl) group and at least two OH (hydroxyl) groups (Fig. 1.9, page 6). These two groups of atoms within the molecule are called **reactive groups** and play important roles in the reactions that take place within cells.

Table 1.1 *Classification and major properties of carbohydrates*

Group	Properties	Examples
monosaccharides general formula: $(CH_2O)_n$ (n = 3 to 9)	small molecules with low molecular mass; sweet tasting; crystalline; readily soluble in water	trioses, e.g. glyceraldehyde ($C_3H_6O_3$); pentoses, e.g. ribose ($C_5H_{10}O_5$); hexoses, e.g. glucose, fructose ($C_6H_{12}O_6$)
disaccharides general formula: $2[(CH_2O)_n] - H_2O$	small molecules with low molecular mass; sweet tasting; crystalline; soluble in water, but less readily than monosaccharides	sucrose, maltose, lactose; all with general formula $C_{12}H_{22}O_{11}$
polysaccharides general formula: $(C_6H_{10}O_5)_n$ (n > 300)	large molecules with high molecular mass; do not taste sweet; not crystalline; insoluble or not readily soluble in water	glycogen, starch, cellulose

DEFINITION

Isomers have the same molecular formula as each other, but different structures. This is due to the different ways in which the atoms or groups are linked within the molecule.

The simplest monosaccharides have three carbon atoms (n= 3) and are called **trioses**. An important triose, **glyceraldehyde**, is formed as an intermediate in the metabolic pathways of respiration and photosynthesis. Its structural formula is shown in Figure 1.7 (see *Extension Material*).

EXTENSION MATERIAL

Glyceraldehyde

Glyceraldehyde has a carbonyl group situated at the end of the molecule at carbon–1 and two hydroxyl groups, one attached to carbon–2 and the other attached to carbon–3. It is known as an aldose, or aldo sugar, because it has an **aldehyde** group, H–C=O. Another triose, dihydroxyacetone, has the same number of carbon, hydrogen and oxygen atoms, but they are arranged differently (Figure 1.8). In this molecule, the carbonyl group is at carbon–2 and there is no hydrogen attached to it. This molecule is known as a ketose, or keto sugar, because it possesses a ketone group, C=O. Its structure is shown in Figure 1.8. These two compounds are isomers – they have the same molecular formula – but they have different structures due to the different linking of atoms or groups within the molecule. This form of isomerism is known as structural isomerism and is common in the carbohydrates.

All the sugars that occur naturally are derived from trioses. All the aldoses are formed from glyceraldehyde and all the ketoses from dihydroxyacetone.

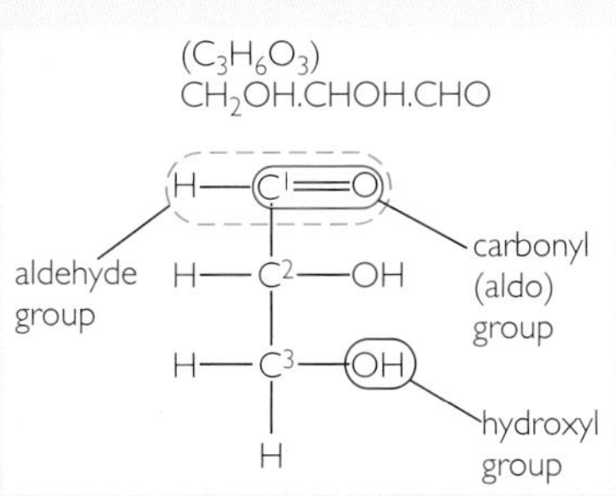

$(C_3H_6O_3)$
$CH_2OH.CHOH.CHO$

aldehyde group
carbonyl (aldo) group
hydroxyl group

Figure 1.7 Structural formula of glyceraldehyde (compare with Figure 1.8).

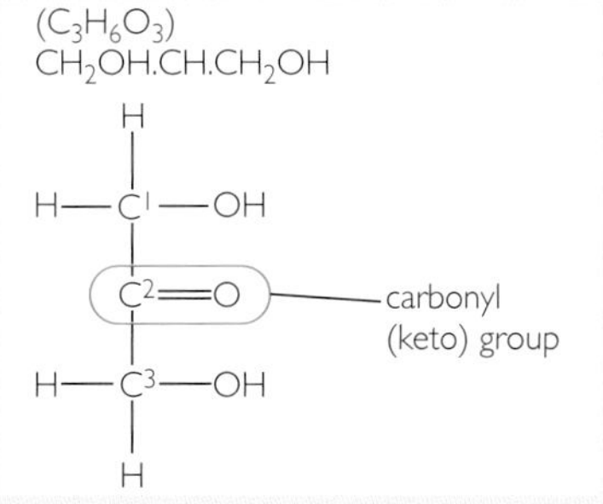

$(C_3H_6O_3)$
$CH_2OH.CH.CH_2OH$

carbonyl (keto) group

Figure 1.8 Structural formula of dihydroxyacetone (compare with Figure 1.7).

Glyceraldehyde illustrates several features of carbohydrates that we have already discussed, such as:

- the ability of carbon atoms to join together to form straight chains
- the property of carbon to form covalent bonds with other atoms
- the ability of carbon to form double bonds, in this case with oxygen (C=O).

Pentoses are monosaccharides with five carbon atoms in the molecule and the general formula $C_5H_{10}O_5$. Like trioses, the pentoses have a carbonyl group and at least two hydroxyl groups. **Ribose**, an important constituent of **RNA** (**ribonucleic acid**) can exist as a chain or in a ring form (Figure 1.9).

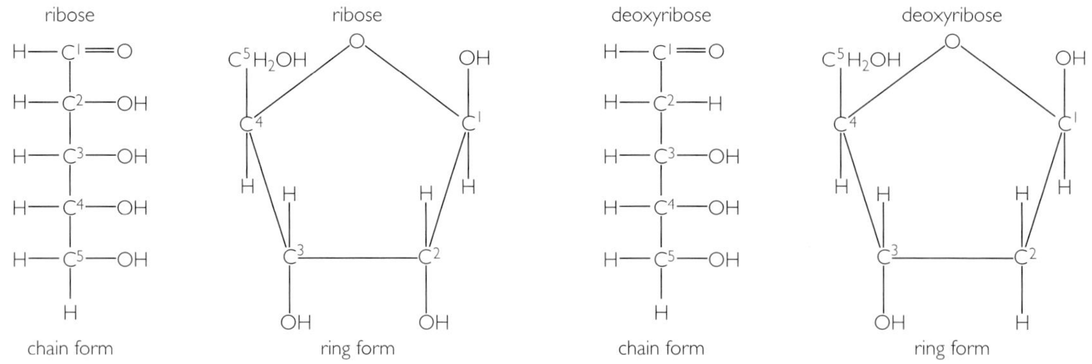

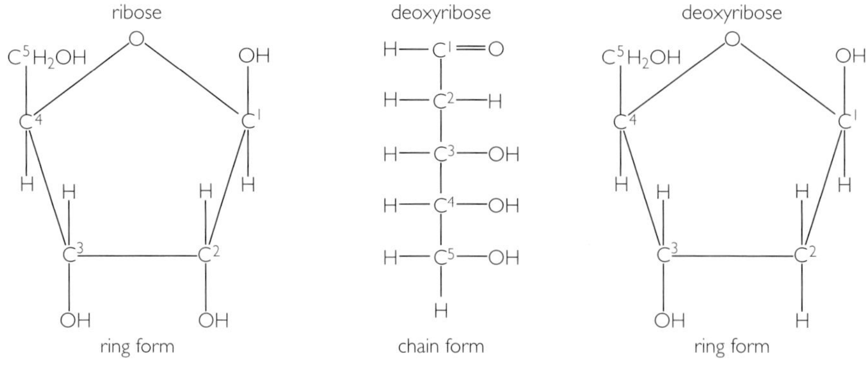

Figure 1.9 Structural formulae of 5-carbon sugar ribose and 5-carbon sugar deoxyribose, showing chain and ring forms.

> **DEFINITION**
>
> **Monomers** are molecules that can be bonded together and become subunits or residues of larger molecules. Two monomers join to form a dimer (e.g. disaccharides, such as maltose) and many monomers join to form polymers (e.g. polysaccharides, such as cellulose).

Deoxyribose is a constituent of **deoxyribonucleic acid** (**DNA**) and is similar to ribose, except that it has one less oxygen atom than ribose.

Hexoses are the six carbon sugars, all having the general formula $C_6H_{12}O_6$. They can exist as straight chains or as rings. When a ring forms, the carbonyl group reacts with one of the hydroxyl groups in the chain. The straight chain and ring forms of two common hexoses, **glucose** and **fructose**, are shown in Figure 1.10.

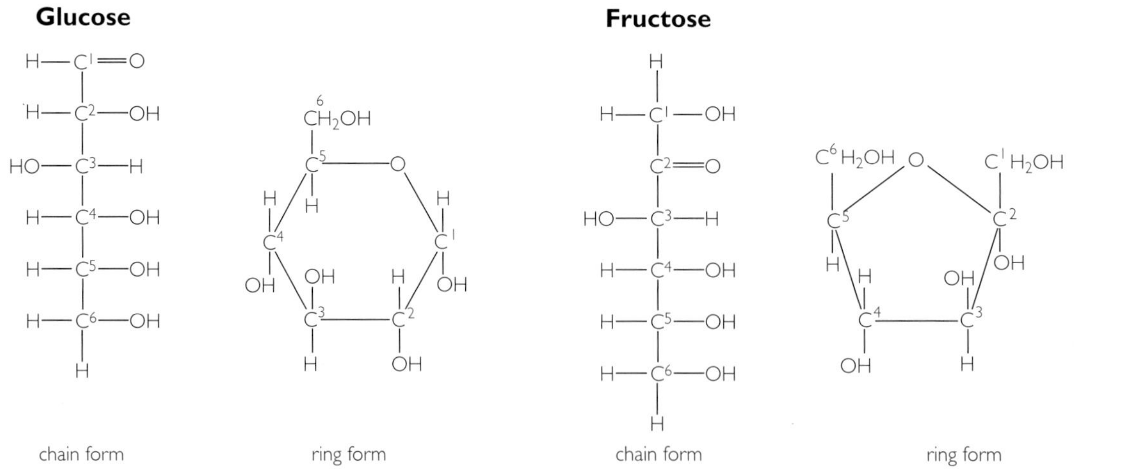

Glucose

chain form

ring form

Fructose

chain form

ring form

Figure 1.10 Structural formulae of 6-carbon sugars, glucose and fructose, showing chain and ring forms.

In glucose, a six-membered ring is formed from five carbon atoms and an oxygen atom. This type of structure is called a **pyranose** ring. Fructose forms a five-membered ring, with four carbon atoms and one oxygen atom, called a **furanose** ring. The six-atom ring form is the more stable ring form for the aldoses (glucose and galactose) and the five-atom ring for the ketoses such as fructose. The ring forms are the more usual forms for both the pentoses and hexoses and it is as such that they are incorporated into disaccharides and polysaccharides.

Glucose can exist in two different ring forms: one where the hydroxyl group on carbon–1 is below the ring (**α-glucose**) and one where the hydroxyl group is above the ring (**β-glucose**) (Figure 1.11). These are known as α- and β-isomers and, because the atoms and groups are arranged differently in space, are examples of **stereoisomerism**. The existence of these two isomers leads to a greater variety in the formation and the properties of polymers. Starch is a polymer of α-glucose and cellulose is a polymer of β-glucose.

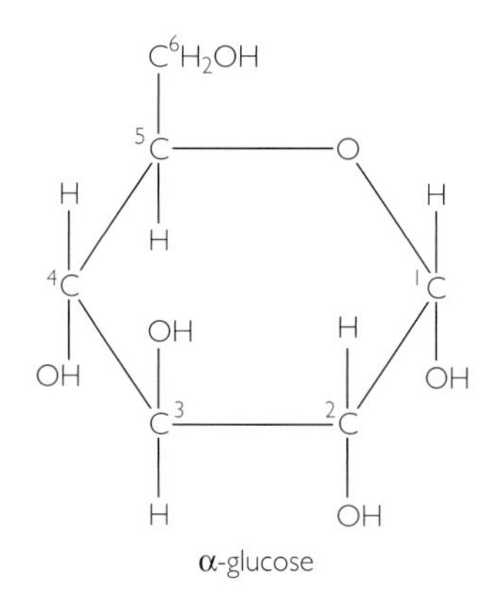

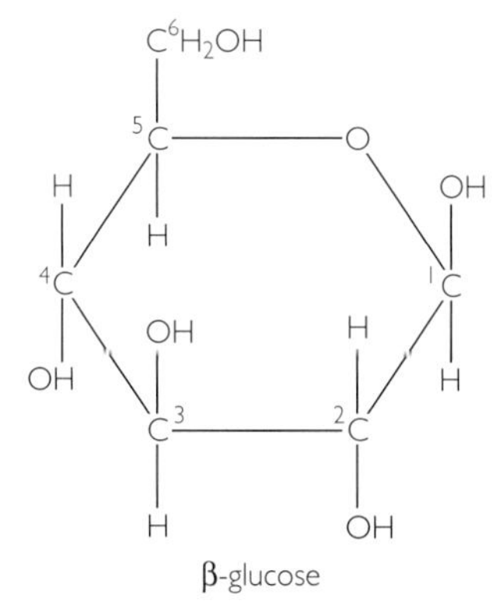

Figure 1.11 Structural formulae of α- and β-glucose, showing stereoisomerism.

Hexoses are sources of energy in respiration and are the building blocks, or monomers, which link together to form disaccharides and polysaccharides. Glucose is the most common respiratory substrate in cells. Glucose is a monomer involved in the formation of maltose, lactose and sucrose. Fructose and galactose are also respiratory substrates and are monomers present in sucrose and lactose respectively.

Disaccharides

When two monosaccharide molecules (monomers) undergo a condensation reaction, a disaccharide molecule is formed and a molecule of water is removed. The bond formed between the two monosaccharide subunits or **residues**, as they are now called, is a **glycosidic bond**. In the example shown in Figure 1.12, two glucose molecules are shown combining to form a molecule of maltose, with the removal of water. Because the bond is formed between the groups attached to carbon–1 of the first glucose molecule and carbon–4 of the second glucose molecule, it is referred to as a 1,4 glycosidic bond.

The formation of sucrose involves a **condensation** reaction between glucose and fructose. In this case, the groups on different carbon atoms are involved and a 1,2 glycosidic bond is formed (Table 1.2).

DEFINITION

A **condensation reaction** between two molecules involves the formation of a bond between two subunits and a water molecule is released (see Fig. 1.12).

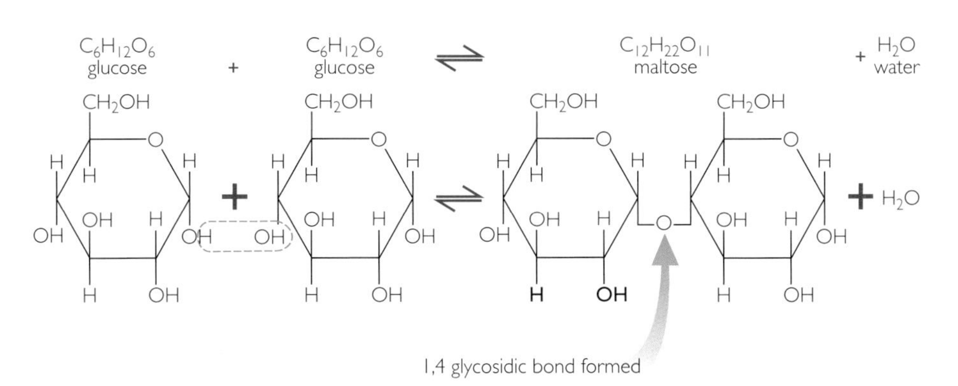

Figure 1.12 Formation of 1,4 glycosidic bond by condensation of two molecules of glucose.

QUESTION

The structural formulae of both glucose and fructose are shown in this chapter. Work out how the 1,2 glycosidic bond is formed.

DEFINITION

A **hydrolysis reaction** involves the breaking of a bond, between the subunits of a large molecule, by the addition of H and OH from a water molecule.

Table 1.2 *Characteristics of the commonly occurring disaccharides*

Disaccharide	Constituent monomers	Type of glycosidic bond	Occurrence and importance
lactose	glucose galactose	1,4	present in mammalian milk, so important in diet of infants
maltose	glucose	1,4	formed by action of amylase (enzyme) on starch during digestion in animals and during germination of seeds
sucrose	glucose fructose	1,2	found in sugar cane and sugar beet; form in which sugars are transported in plants; storage compound in some plants, e.g. onions

On **hydrolysis**, which requires water to be present, disaccharides can be split into their constituent monosaccharides. Within cells, these reactions are catalysed by specific enzymes. In the laboratory, it is possible to hydrolyse disaccharides by heating in solution with acids.
(For details of condensation and hydrolysis reactions, see *Appendix: Physical science background.*)

EXTENSION MATERIAL

Reducing and non-reducing sugars

All the monosaccharides are described as **reducing sugars**. When solutions of these sugars are heated with Benedict's reagent (an alkaline solution of blue copper sulphate [$CuSO_4$]), a red-brown precipitate of insoluble copper oxide [Cu_2O] is formed. The aldehyde groups of aldoses, such as glucose and galactose, are able to reduce the copper(II) [Cu^{++}] in the copper sulphate to copper(I) [Cu^{+}] in copper oxide. As a result, the aldehyde groups are oxidised to **carboxyl (COOH)** groups. In sugars that do not have aldehyde groups, such as the ketoses (fructose), the carbonyl group on carbon–2 changes place with the hydroxyl group on carbon–1, so they can also reduce copper(II) to copper(I) when heated with the reagent.

In some disaccharides, for example maltose and lactose, an aldehyde group is available, so a positive result is obtained when these sugars are tested with Benedict's reagent. When sucrose is heated with the reagent, no colour change occurs, as the aldehyde groups are not available. Sucrose is described as a **non-reducing sugar**. If a solution of sucrose is boiled with dilute hydrochloric acid, cooled, neutralised and retested with the reagent, the familiar brown-red precipitate develops, because the sucrose has been hydrolysed into its constituent monosaccharides: glucose and fructose. The Benedict's test is frequently used in the identification of sugars. (See *Practicals: Quantitative estimation of sugars and Identification of food constituents in milk.*)

Polysaccharides

Polysaccharides are **macromolecules**, with very large relative molecular masses ranging from 5000 to 10 000. They are **polymers** formed from large numbers of monosaccharide **monomers**, joined together by covalent bonds by a process known as **condensation polymerisation**. This process is essentially similar to the way in which two monosaccharides are joined to form a disaccharide, which has already been described.

The commonly occurring polysaccharides starch, glycogen and cellulose are all polymers of glucose. The glucose monomers are linked together by glycosidic bonds. The different nature of these polysaccharides depends on the isomer of glucose involved and on the type of glycosidic bond (Table 1.3).

Table 1.3 *Summary of features of some common polysaccharides*

Polysaccharide	Monomer	Type of glycosidic bond	Shape of molecule
starch (amylose)	α-glucose	α-1,4	unbranched chain wound into a helix
starch (amylopectin)	α-glucose	α-1,4 with some α-1,6	tightly packed branched chain
glycogen	α-glucose	α-1,4 with more α-1,6 than amylopectin	very branched compact molecule
cellulose	β-glucose	β-1,4	unbranched straight chains

Starch

Starch is a polymer of α-glucose monomers and is a mixture of amylose and amylopectin (Figure 1.13). Amylose makes up about 30 per cent of starch and consists of unbranched chains in which the monomers are joined by α-1,4 glycosidic bonds. The molecules usually contain more than 300 glucose monomers and adopt a helical shape. The coils have six monomers per turn and are held together by hydrogen bonds formed between the groups attached to the carbon atoms. When iodine in potassium iodide solution is used to

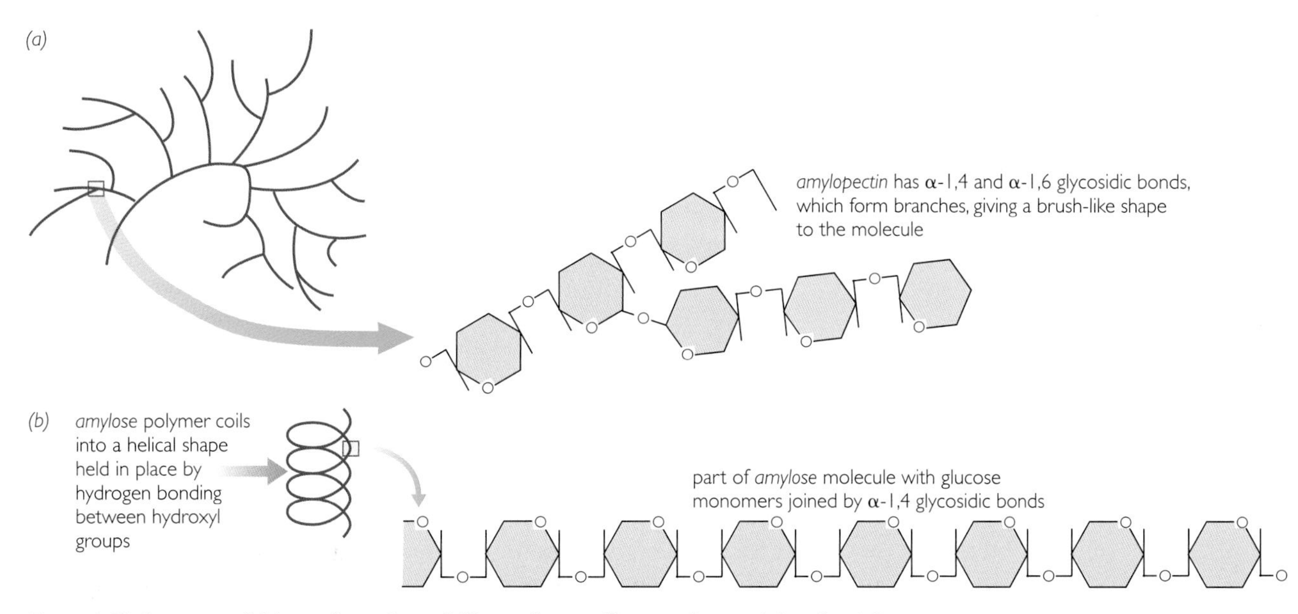

***Figure 1.13** Structure of (a) amylopectin and (b) amylose to illustrate how α-1,4 and α-1,6 glycosidic bonds affect the three-dimensional shape of polysaccharide.*

detect the presence of starch, a complex forms with the glucose monomers in the amylose helix. This leads to the formation of the characteristic blue-black colour. Amylopectin, which constitutes the remaining 70 per cent of starch, consists of chains of glucose monomers linked with α-1,4 glycosidic bonds. Branches arise from these chains due to the formation of α-1,6 bonds at various points, about every 20 to 30 residues along their length. The resulting molecule, consisting of several thousand monomers, is much branched and also coiled into a compact shape.

Starch functions as an important storage molecule in plants. It is particularly well suited to this function because:

- it is compact and does not take up much space
- it is insoluble so it cannot move out of the cells in which it is stored
- it has no osmotic effects
- it does not become involved in chemical reactions in the cells
- it is easily hydrolysed to soluble sugars by enzyme action when required.

Figure 1.14 Photomicrograph of starch grains from a potato.

Starch molecules are built up into starch grains inside special structures called **amyloplasts**, which are present in the cytoplasm of plant cells. Starch grains may also be built up in the stroma of chloroplasts (see Chapter 4). The starch molecules appear to be deposited in concentric layers around a central point, forming the grains, which have characteristic shapes according to the species (Figure 1.14).

QUESTION

Amylopectin and glycogen are branched polysaccharides. What are the similarities and differences between these two molecules?

Glycogen

Glycogen is a polymer of α-glucose monomers in which there are both α-1,4 and α-1,6 glycosidic bonds (Figure 1.15a). It is very similar to amylopectin in that it is branched, but the branch points occur more frequently, every 8 to 12 residues, forming a very compact structure. Glycogen is an energy storage molecule in animals, where it occurs in the liver cells and in muscle tissue (Figure 1.15b). It is also found in the cytoplasm of bacterial cells. As with amylose and amylopectin in starch, glycogen is well suited to its function, taking up little space and preventing too high a concentration of glucose in the cells. It can be readily hydrolysed to glucose for use as a respiratory substrate when required.

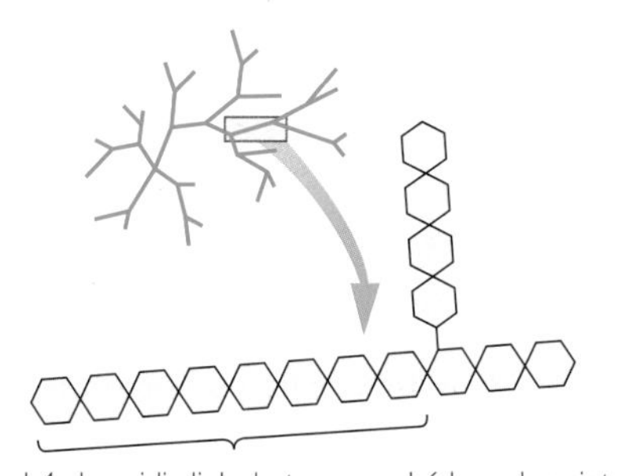

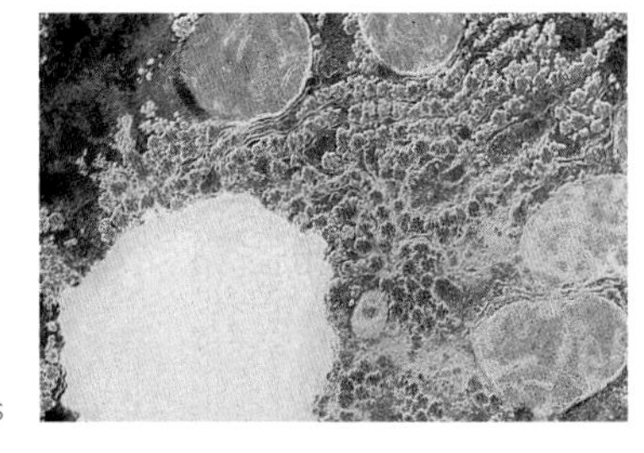

Figure 1.15 (a) Structure of glycogen to illustrate a branched polysaccharide of animal origin; (b) electronmicrograph of glycogen granules (yellow and green) in cytoplasm of liver cell, showing association with endoplasmic reticulum (pink) (× 12 000).

Cellulose

Cellulose is a polymer of *β*-glucose monomers joined by *β*-1,4 glycosidic bonds to form straight unbranched chains (Figure 1.16). Due to the orientation of the monomers, there is no tendency for the molecules to coil into a helical shape. Each chain contains thousands of *β*-glucose residues with hydroxyl groups projecting out all round. Hydrogen bonding occurs between the hydroxyl groups on adjacent chains forming cross links that hold the chains together. Up to 2000 such chains can be held together to form a **microfibril**, which can be many micrometres in length. Microfibrils have great tensile strength, enabling them to resist pulling forces.

Cellulose is an important structural component of plant cell walls, where its tensile strength is important. The cellulose microfibrils are embedded in a kind of 'cement', or **matrix**, which holds them together. The matrix consists of a mixture of pectins and hemicelluloses (see also Chapter 4).

Cellulose can be hydrolysed in the laboratory only by treatment with concentrated acids. Despite the abundance of cellulose, there are relatively few groups of living organisms capable of producing the enzyme **cellulase**, which catalyses the digestion of cellulose to glucose. Some prokaryotes and fungi are able to synthesise the enzyme, but the ruminant mammals, for whom cellulose is such an important food source, depend on large populations of symbiotic bacteria in their guts to break down the cellulose for them. These mammals can then absorb the nutrients released by the bacteria.

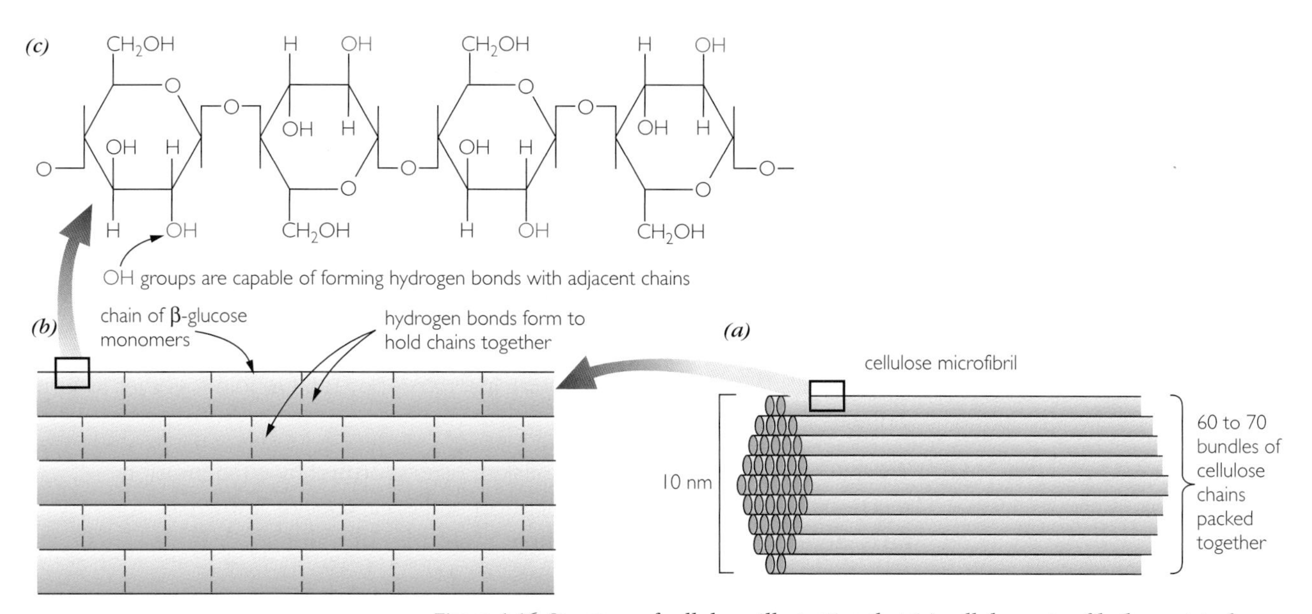

Figure 1.16 Structure of cellulose, illustrating that (a) cellulose microfibrils consist of (b) hydrogen-bonded chains of (c) β-glucose monomers.

EXTENSION MATERIAL

Other polysaccharides

Other polysaccharides and some closely related compounds are of biological significance and any survey of the carbohydrates would be incomplete without reference to them.

- **Callose** is polymer of glucose found in plants. The monomers are linked by 1,3 glycosidic bonds. Callose is found lining the pores of the sieve plates in phloem sieve tubes and should not be confused with *callus*, a type of parenchyma tissue often produced near the surface of a wound or in the formation of a union between the scion and the stock in grafting.
- **Inulin** is a polymer of fructose, found as a storage carbohydrate in some groups of plants, for example *Dahlia* tubers.
- **Pectins** and hemicelluloses are associated with the cellulose cell walls in plants. Pectins are polysaccharides formed from galactose and organic acid. The chains have negative charges and attract calcium (Ca^{2+}) ions, forming calcium pectate, which attracts and holds water molecules, forming a gel. This gel makes up the matrix of the cell wall in which the cellulose microfibrils are embedded. Hemicelluloses are very similar in composition to cellulose in that they consist of short chains of β-glucose residues linked by β-1,4 glycosidic bonds, but they have in addition side branches of other sugars, such as galactose. The hemicellulose molecules are held tightly by hydrogen bonds to the cellulose microfibrils in the cell wall and also to the pectate molecules, thus linking the pectate and the cellulose.
- **Chitin** has a structure very similar to cellulose, in that bundles of long parallel chains are formed. Each chain is composed of β-glucose residues linked by β-1,4 glycosidic bonds, but the hydroxyl group at carbon–2 of the β-glucose has been replaced by $NH.CO.CH_3$. These residues are often referred to as amino sugars. Chitin occurs in the walls of fungal hyphae and forms the exoskeleton in arthropods (Figure 1.17).
- **Murein** is a polysaccharide present in the cell walls of prokaryotes, where the chains are composed of the same amino sugar found in chitin molecules, together with another similar amino sugar. These two alternate along the chain. Adjacent chains are cross-linked by amino acids.
- **Mucopolysaccharides**, found in bone, cartilage and the synovial fluid within joints, are polymers of organic acids derived from sugars and amino acids.

Figure 1.17 Tropical stag beetle exoskeleton, composed of chitin, a hard polymer of amino sugars.

Lipids

Lipids are **fats** and **oils**, organic compounds containing carbon, hydrogen and oxygen. The same three elements are involved in the structure of carbohydrates, but in lipids the amount of oxygen in the molecule is much less than in the carbohydrates. Lipids are insoluble in water, but soluble in organic solvents such as acetone and ether. They are relatively small molecules compared to the polysaccharides, but because they are insoluble they tend to join together to form globules.

The naturally occurring fats and oils are esters formed by condensation reactions between glycerol (an alcohol) and organic acids known as **fatty acids**. During formation, three molecules of water are removed.

$$\underset{\text{(an alcohol)}}{\text{glycerol}} + \underset{\text{(organic acids)}}{\text{fatty acids}} \rightleftharpoons \underset{\text{(ester)}}{\text{lipid}} + \text{water}$$

Glycerol, with the formula $C_3H_8O_3$, has three **hydroxyl (OH)** groups, all of which can take part in condensation reactions with a fatty acid. The resulting ester is called a **triglyceride** (or triacylglycerol) (Figure 1.18).

DEFINITION

A **triglyceride** is a lipid formed from the condensation reaction between a glycerol molecule and three fatty acid molecules.

Fatty acid molecules are much larger than glycerol molecules and consist of long, non-polar hydrocarbon chains with a polar, carboxyl group (COOH) at one end. There are many different fatty acids present in living organisms, but they can be divided into saturated fatty acids, which possess only single bonds in their hydrocarbon chains, and unsaturated fatty acids, which have one or more double bonds. Lipids composed of saturated fatty acids are termed **saturated fats** and those built up of unsaturated fatty acids are **unsaturated fats**. In order to appreciate the differences between the two, let us compare the structure of stearic acid, with a formula of $C_{17}H_{35}COOH$, with oleic acid, $C_{17}H_{33}COOH$ (Figure 1.19). Without looking at the arrangement of the atoms within the molecules, it can be seen that oleic acid has two fewer hydrogen atoms than stearic.

Fatty acids play an important role in cells. They can be broken down and oxidised to release energy for cell metabolism or built up into triglycerides, forming an energy store. Fatty acids can also be converted to phospholipids, which are important constituents of cell membranes.

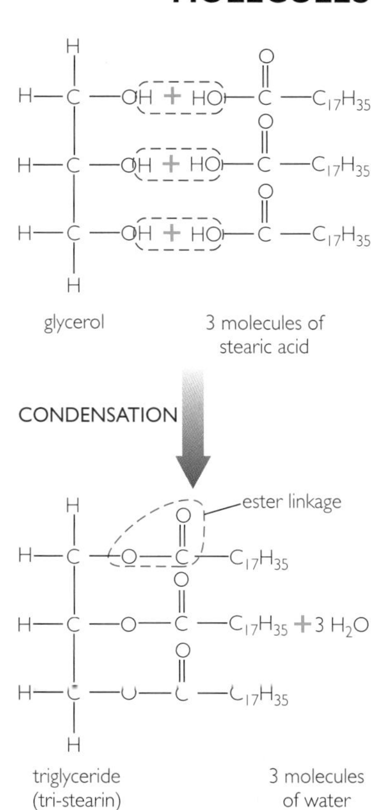

Figure 1.18 Formation of a triglyceride by condensation of three molecules of fatty acid with one molecule of glycerol.

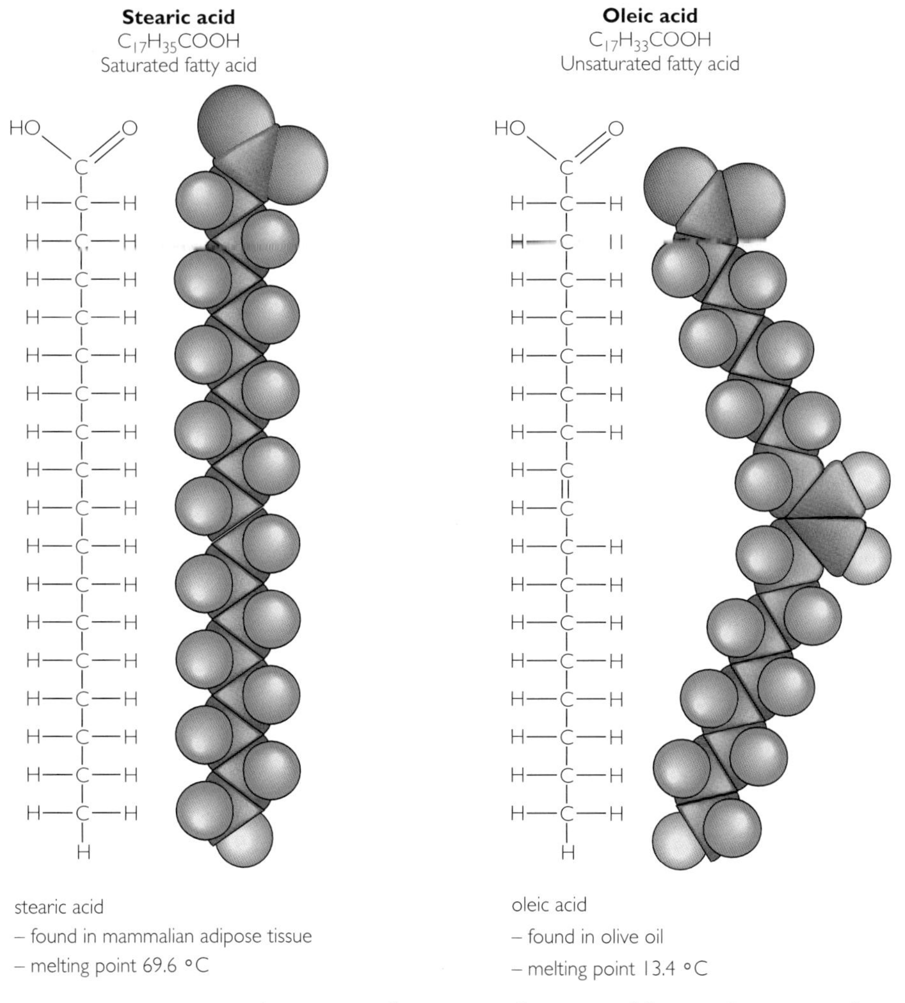

Figure 1.19 Comparison of structure and properties of a saturated fatty acid (stearic acid) and an unsaturated fatty acid (oleic acid). The right-hand section of each diagram shows a molecular model of the substance.

Definition

Saturated fats are made up of saturated fatty acids which have:

- no C=C double bonds in their hydrocarbon chains
- higher melting points, so that they are solid at room temperature.

Unsaturated fats consist of unsaturated fatty acids which have:

- one (monounsaturated) or more (polyunsaturated) C=C double bonds in their hydrocarbon chains
- lower melting points so are liquid at room temperature.

Saturated fats are more readily converted to cholesterol in the human body than unsaturated fats.

Extension Material

Metabolic water

When organic molecules are oxidised, water and carbon dioxide are released. The water is referred to as **metabolic water** and is of great importance to animals in a terrestrial environment. The oxidation of 1 g of glucose yields 0.6 g of metabolic water, and the oxidation of 1 g of protein yields slightly less (0.5 g), but the oxidation of 1 g of fat gives 1.07 g.

> **QUESTION**
>
> Explain why lipids are better storage molecules than starch or glycogen.

Triglycerides are the commonest lipids in living organisms and their primary importance is as energy stores. They are compact and insoluble and can be stored at high concentrations in cells, where they occur as small droplets suspended in the cytoplasm. They release approximately twice as much energy per gram as carbohydrates, but the process of energy release requires oxygen. In addition, more water is released on oxidation than from the oxidation of carbohydrates. This is known as **metabolic water** and is important to organisms living in dry climates. Oils are the major food store in many seeds (sunflower, rape) and fruits (palm, olive). In animals, fats are stored in **adipose** tissue, which consists of large fat cells found below the skin and around the body organs. In addition to providing an energy store, the fatty tissues protect vital organs and provide insulation and, in aquatic animals, buoyancy.

Waxes are esters of fatty acids and long-chain alcohols. Their main function in plants and animals is waterproofing. They form an additional layer on the cuticle of the epidermis of the leaves, fruit and seeds of some plants, where they prevent the entry or evaporation of water. In animals, waxes occur in the skin, fur and feathers of vertebrates and there is often a layer of wax deposited on the outside of the cuticle in insects. Worker bees are able to manufacture wax, which they use in the construction of the honeycomb and the cells, in which the eggs are laid and where the development of the larvae takes place.

> **DEFINITION**
>
> If a part of a molecule is **hydrophilic** then it can associate with water molecules. If it is **hydrophobic** then it tends to move away from water molecules.

Phospholipids

Phospholipids are similar to lipids in that they are esters of fatty acids and glycerol, but one of the fatty acid chains is replaced by a polar group associated with a phosphate molecule (Figure 1.20). The polar group is very soluble in water (hydrophilic), whereas the non-polar hydrocarbon chains of the fatty acids are not (hydrophobic). At an air–water, or oil–water interface, the phospholipid molecules orientate themselves so that their polar heads are in the water. Phospholipids are important constituents of cell membranes and their involvement is described in Chapter 4.

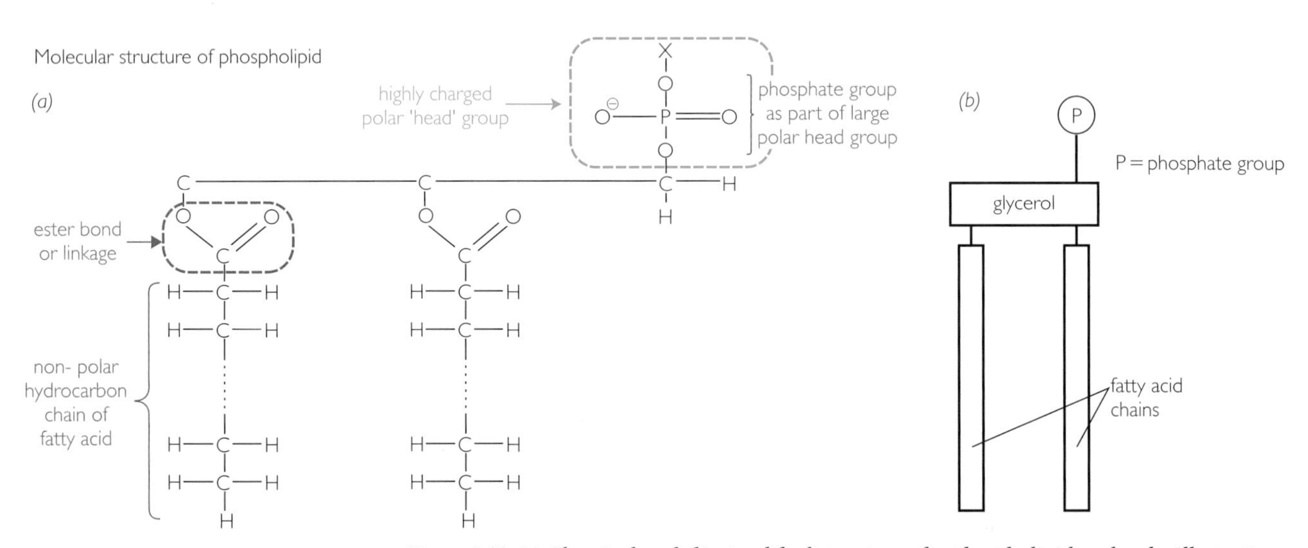

Figure 1.20 (a) Chemical and (b) simplified structure of a phospholipid molecule, illustrating how its polar nature derives from a highly charged phosphate group (and other associated groups) at the 'head' of the molecule. The polar phosphate group is hydrophilic (water-loving). The non-polar hydrocarbon chains are hydrophobic (water-hating).

Proteins

Proteins are complex organic molecules containing carbon, hydrogen, oxygen and nitrogen. Sometimes they contain sulphur and may form complexes with other molecules containing phosphorus, iron, zinc or copper. Proteins are macromolecules, with relative molecular masses between 10^4 and 10^6. They consist of one or more unbranched **polypeptide** chains built up from **amino acid** monomers linked by **peptide bonds** (see Figure 1.22).

Each protein has a characteristic three-dimensional shape resulting from the coiling and folding of the constituent polypeptide. When considering the structure of a protein, it is usual to describe three or four different levels of organisation of the molecule. These levels are often referred to as the primary, secondary, tertiary and quaternary structures.

Primary structure

The **primary structure** of a protein is the number, type and sequence of the amino acids that make up the polypeptide chain. The primary structure is specific to each protein and coded for by the DNA of the cell in which the protein is made. As the amino acids are fundamental to the structure of proteins, it is worth considering their structure and properties and the way in which they are bonded together.

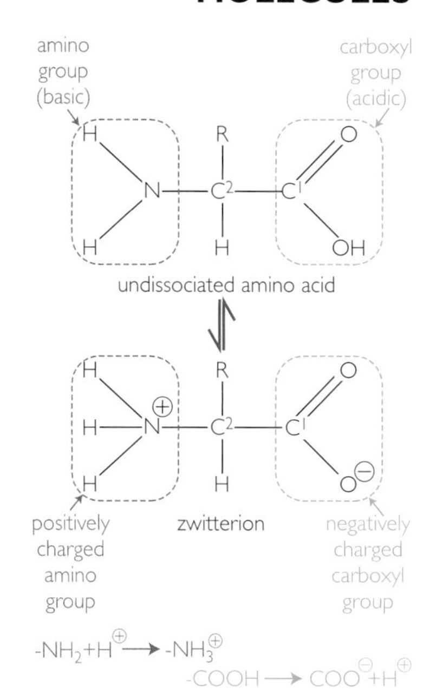

Figure 1.21 Structure of undissociated and ionised (zwitterion) forms of an amino acid, to show basic and acidic properties.

Definition

The **primary structure** of a protein is the number, type and sequence of amino acids held together by peptide bonds.

Structure of amino acids: Amino acids are crystalline solids and are soluble in water. They have the general formula $NH_2.RCH.COOH$, where R varies from a single hydrogen atom to more complex groups, including ring structures. They contain an amino or amine (NH_2) group at one end of the molecule and a carboxylic (COOH) group at the other end. There are 20 common amino acids found in living organisms.

The amino group has basic properties and the carboxyl group has acidic properties. In an amino acid, the –COOH group (acidic) dissociates, releasing H^+ ions and the $-NH_2$ group (basic) has an affinity for and can combine with H^+ ions. Compounds with both basic and acidic properties are called **amphoteric**. In living cells, the pH is usually neutral, so both the groups become ionised, leaving the molecule positively charged at one end and negatively charged at the other (Figure 1.21). This type of molecule is sometimes referred to as a **zwitterion**, or double ion.

Two amino acids can join together in a condensation reaction to form a **dipeptide**. The bond formed between the two is called a peptide bond and a molecule of water is removed. Further condensation reactions enable additional amino acids to be added, resulting in a polypeptide chain (Figure 1.22).

The most important role of amino acids in cells is as the monomers involved in protein synthesis. Green plants are able to synthesise all the amino acids they require from the products of photosynthesis and nitrate ions absorbed from the soil. Animals can synthesise some of the amino acids they require, but need to obtain the rest from their diet. We need to obtain eight amino acids from our diet. These are known as the essential amino acids, and we are able to synthesise all the rest that we require to make the proteins we need. Amino acids are also involved in the synthesis of other compounds, such as nucleic acids and cytochromes.

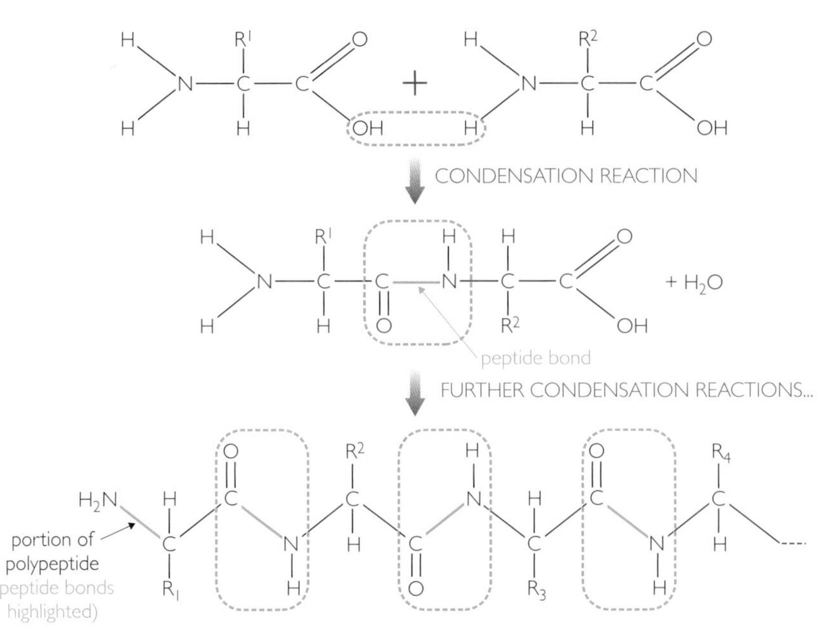

Figure 1.22 Formation of a peptide bond by condensation of two amino acid molecules, and further condensation to generate a polypeptide. R^1 and R^2 are the amino acid residues.

DEFINITION

The **secondary structure** of a protein is brought about by the formation of hydrogen bonds between the CO and NH groups of the amino acids in the polypeptide chain. This produces α-helices and β-pleated sheets.

Secondary structure

In protein molecules, polypeptide chains are either coiled into a spiral spring, the ***α*-helix**, or linked together to form ***β*-pleated** sheets. Both these arrangements of the chains constitute the secondary structure of the protein. These are stable structures, maintained by hydrogen bonding between different groups on the amino acid residues of the chains.

In an *α*-helix, the coils in the polypeptide chain are held in place by hydrogen bonds that form between the hydrogen atoms of the NH group of one amino acid and the oxygen atom of the CO group of another amino acid further along the chain. It has been estimated that the *α*-helix makes one complete turn every 3.6 amino acid residues. In the fibrous protein keratin, which is found in hair, nails and skin, several *α*-helices are held together by bonds formed between adjacent chains and this protein does not have a tertiary structure. The bonds, in this case disulphide bridges, form cross-links and the bundles of molecules have strength and the ability to stretch (Figure 1.24).

The fibrous protein collagen forms insoluble fibres that have a high tensile strength. Collagen is the major fibrous protein in skin, bone, tendon and cartilage. It consists of three polypeptide chains coiled around each other, forming a triple helix, or **superhelix** (see Figure 1.23). The polypeptide chains, consisting mainly of glycine and proline residues, do not form *α*-helices because the proline residues lack the NH groups needed for internal hydrogen bond formation. The three chains are held together by hydrogen bonding between the NH groups of the glycine residues of one chain and the CO groups of the proline residues of an adjacent chain. A number of superhelices associate to form collagen fibres further stabilised by covalent cross-links. Collagen is flexible but does not stretch.

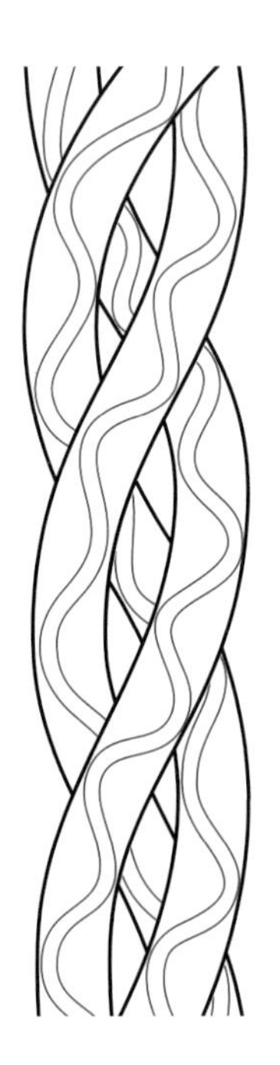

Figure 1.23 Collagen superhelix of three polypeptide chains.

In a *β*-pleated sheet, hydrogen bond formation takes place between the CO and NH groups of the amino acid residues of one chain and the NH and CO groups of neighbouring chains. This arrangement, giving the protein a high

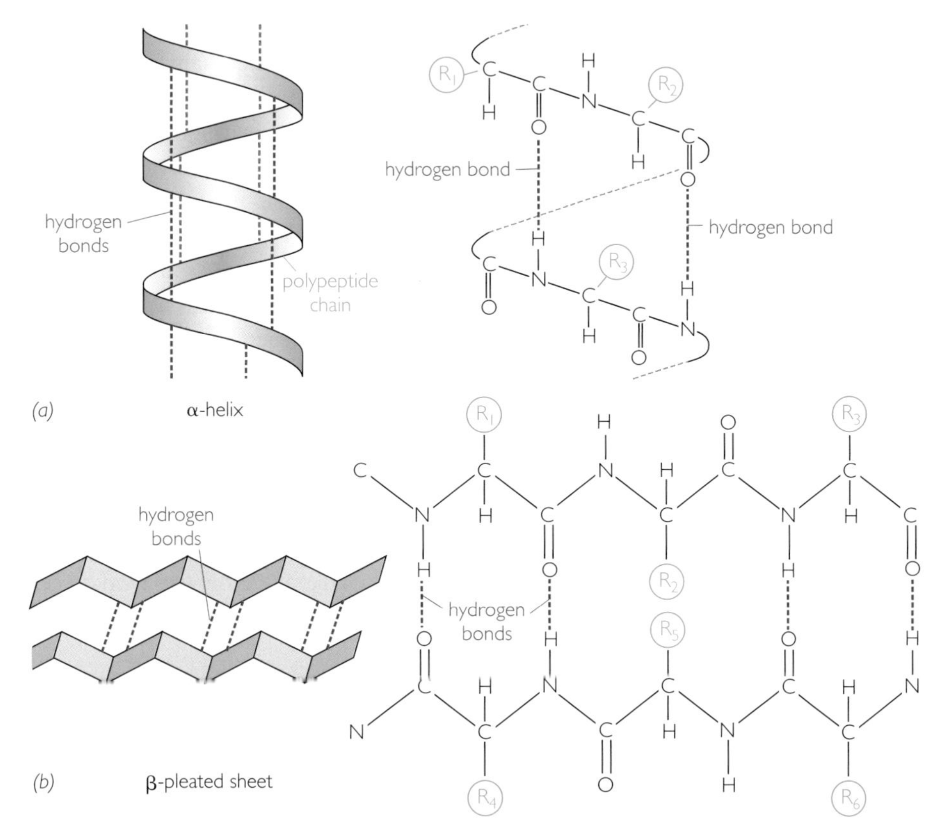

Figure 1.24 Comparison between the arrangement of the amino acids in (a) an α-helix and (b) a β-pleated sheet, illustrating the role of hydrogen bonding in secondary protein structure.

tensile strength, is found in silk fibroin, produced by silkworms. The fibroin cannot be stretched, but is very supple.

> **DEFINITION**
>
> The **tertiary structure** of a protein is the arrangement of the **α-helices** and **β-pleated** sheets of a polypeptide chain forming a three-dimensional structure held together by hydrogen, covalent and ionic bonds.

Tertiary structure

In most proteins there are regions of α-helix and regions of β-pleating, but the folding of the polypeptide chain into a compact, globular shape is referred to as the **tertiary structure**. The bending and folding is irregular and it is the result of the formation of different types of bonds between the amino acid residues.

The nature of the R-groups of the constituent amino acids and their interactions play an important role in determining and maintaining the specific shape of a protein molecule. Some of the R-groups may be polar and attracted to water, whilst others are non-polar. The polar groups are referred to as hydrophilic, or 'water-loving', and the non-polar groups are hydrophobic, or 'water-hating'. Folding of the polypeptide chain results in a structure with most of the hydrophilic groups on the outside of the molecule and the hydrophobic groups on the inside. This arrangement is more stable, particularly for globular proteins in living cells.

Some of the R-groups can form:

- hydrogen bonds between their hydrogen and oxygen atoms and those of other R-groups
- ionic bonds
- disulphide bridges (applies to sulphur-containing amino acid residues).

The different types of bonds are illustrated in Figure 1.25.

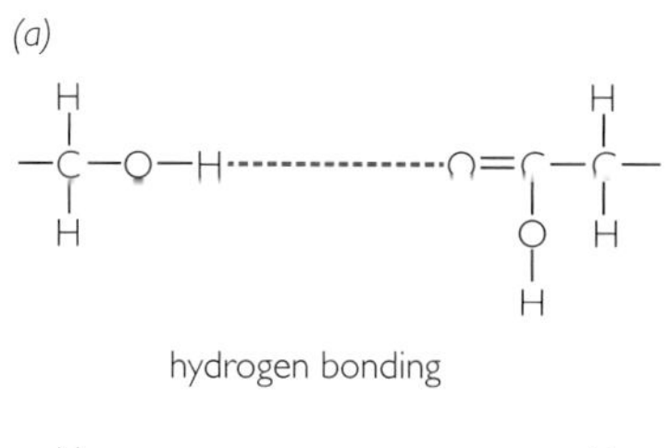

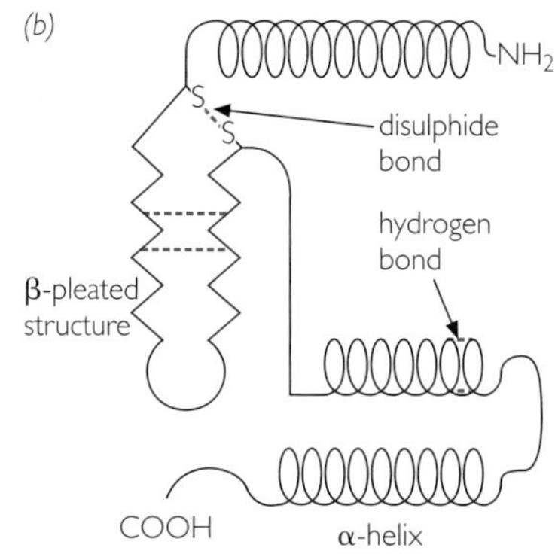

Figure 1.25 (a) Types of bonding commonly found in the tertiary structure of proteins. (b) Within the tertiary structure of a globular protein, there are regions of α-helix and of β-pleating, as shown. The different types of bonding are also illustrated.

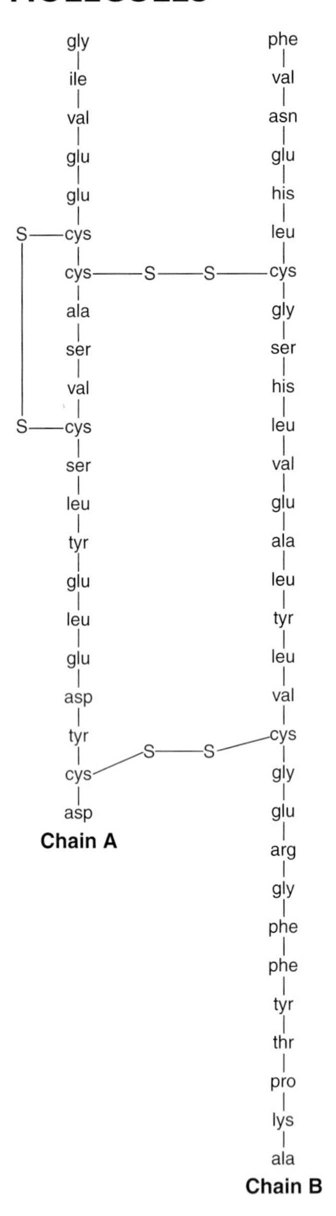

Figure 1.26 The insulin molecule consists of 51 amino acids in two polypeptide chains, chain A and chain B, held together by two disulphide (–S–S–) bridges.

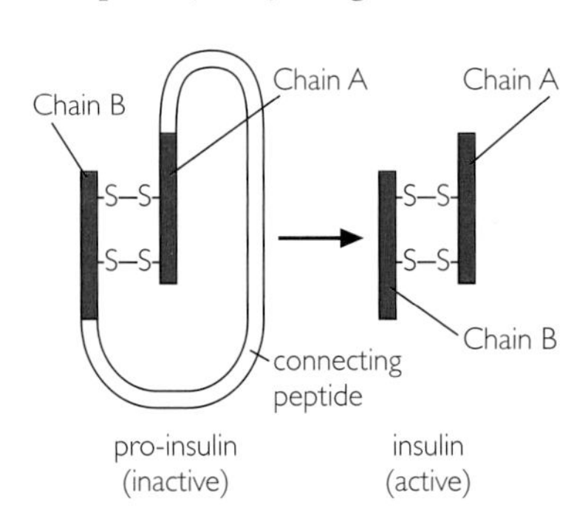

Figure 1.27 Conversion of pro-insulin to insulin.

Quaternary structure

Complex proteins may consist of more than one polypeptide chain and so are described as having a **quaternary structure**. The polypeptide chains in complex proteins may be all of one type or of different types (Figure 1.26). The hormone **insulin** is a small globular protein, consisting of two short polypeptide chains held together by disulphide bridges (Figure 1.26). Chain A has 21 amino acid residues and Chain B has 30 amino acid residues. Insulin is produced in the β-cells of the islets of Langerhans in the pancreas and is involved with the control of the blood glucose level. It is produced as an inactive precursor, **pro-insulin**, and is converted into its active form by removal of parts of its polypeptide chain (Figure 1.27).

In haemoglobin, a protein present in the blood, there are four polypeptide chains held together by bonds of the types described above. Two of the chains, the α-globin chains, each contain 141 amino acid residues, and the other two, the β-globin chains, contain 146 amino acid residues. Each chain is associated with an iron-containing haem group. The structure of haemoglobin was worked out by Kendrew and Perutz in 1959 (Figure 1.28).

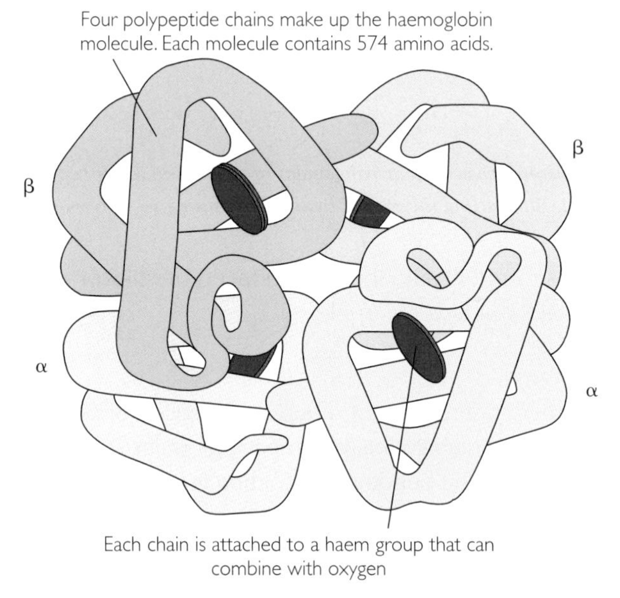

Figure 1.28 Structure of a molecule of haemoglobin, showing how the α– and β– chains associate around the iron-containing haem groups in the quaternary structure of a globular protein.

Proteins are abundant in living organisms, but because of their diverse nature it is difficult to produce a simple classification. It is customary to group them either according to their structure (Table 1.4) or to their functions (see *Extension Material*, page 19) within living organisms.

Under certain circumstances, the three-dimensional shape of a globular protein molecule can change. This change may be temporary or permanent, affecting the tertiary structure of the protein, but not altering its primary structure. Alteration of the tertiary structure affects the biological role of the protein, especially in the case of enzymes (see Chapter 3), which depend on the specific configuration of the active site in order to form an enzyme–substrate complex. This loss of shape is referred to as **denaturation** and can be caused by an increase in temperature, a change in pH, high concentrations of salts, the presence of ions of heavy metals and organic solvents.

Table 1.4 *Classification of proteins according to their structure*

Fibrous	Globular
secondary structure important; consist mainly of α-helix or β-pleated sheets	tertiary structure important; bent and folded into spherical shapes
insoluble in water	soluble in water
structural functions e.g. keratin, collagen	act as enzymes, antibodies, hormones e.g. amylase, globulins, insulin

DEFINITION

Denaturation of a protein involves the disruption of the bonds that hold the tertiary structure in place. This can be brought about by:

- heat energy breaking hydrogen bonds – many but not all enzymes are denatured at temperatures above 50 °C
- extreme changes in pH – extremely acid or extremely alkaline conditions alter the charges on the amino acid side chains, thus breaking ionic bonds.

DEFINITION

A protein that has a **quaternary structure** is composed of at least two polypeptide chains held together by hydrogen, covalent and ionic bonds.

EXTENSION MATERIAL

Table 1.5 *Classification of proteins according to their function*

Function	Example	Location
contraction	myosin, actin	muscle tissue
enzymes	amylase	endosperm of starchy seeds; human duodenum
hormones	insulin	present in blood; secreted from islets of Langerhans in pancreas
homeostasis	soluble proteins	can act as buffers, stabilising pH in body cells
protection	antibodies fibrinogen	produced by lymphocytes in the blood present in blood as part of blood clotting mechanism
storage	aleurone layer	in seeds
structure	collagen keratin	skin hair
transport	haemoglobin	transport of oxygen in vertebrate blood

PRACTICAL

Quantitative estimation of sugars

Introduction

Concentrations of reducing sugars can be determined semi-quantitatively using Benedict's reagent and a range of colour standards. Quantitative estimations of glucose concentrations may be determined conveniently using suitable test strips, such as Diabur 5000. The concentration of sucrose, a non-reducing sugar, can be estimated by first adding a drop of 10% invertase (sucrase) concentrate to 2 cm^3 of the solution to be tested and leaving for 30 minutes at room temperature. After enzyme treatment, the solution is tested for the presence of a reducing sugar. This method is preferable to acid hydrolysis.

Materials

- Range of food samples to be tested
- Mortar and pestle
- Beaker to use as boiling water bath
- Test tubes
- Diabur 5000 reagent strips
- Benedict's reagent
- Pipettes or syringes
- 10% invertase (sucrase) concentrate
- Standard glucose solutions: 2.0, 1.0, 0.5, 0.1, 0.05, 0.02 and 0.01 per cent.

WEAR EYE PROTECTION

Method

1 To produce a range of colour standards, use a series of glucose solutions of known concentration. Add 3 cm^3 of each of these solutions to a series of appropriately labelled test tubes, each containing 5.0 cm^3 of Benedict's reagent. These test tubes should then be placed in a boiling water bath for 8 minutes, then left to cool in air.

2 To estimate the concentration of reducing sugars in the food samples, pipette 5.0 cm^3 of Benedict's reagent into a test tube and add 3 cm^3 of the solution to be tested. Heat in a boiling water bath for 8 minutes, leave to cool, then compare the colour produced with the colour standards.

3 If using Diabur test strips, a strip should be dipped into the solution to be tested, removed, and the colours produced compared with the colour chart after two minutes. This method is specific for glucose, and gives quantitative results.

4 You can also estimate reducing sugar content by carrying out the Benedict's test, then removing the precipitate by filtration. Dry the precipitate and record the mass. Provided that there is an excess of Benedict's reagent the mass of the precipitate is proportional to the concentration of reducing sugars.

Results and discussion

1 Tabulate all your results suitably.

2 Present the results in a suitable graphical form.

3 What are the sources of error in this experiment? How could it be improved?

Further work

1 Estimate the reducing sugar content of a range of fruit juices, both fresh and packaged.

2 Investigate the changes in reducing sugar content during, for example, the ripening of fruit.

3 Find out about other quantitative methods for the determination of reducing sugar content.

PRACTICAL

Identification of food constituents in milk

Introduction

The purpose of this practical is to identify and, where possible, quantify the food constituents of milk. There are a number of possibilities for comparing the content of different types of milk, and of milk treated in different ways, such as pasteurised, sterilised and UHT.

Materials

- Samples of different types of milk
- Benedict's reagent
- Biuret reagent (Dissolve 8 g of sodium hydroxide in 800 cm^3 of distilled water. Add 45 g of sodium potassium tartrate and dissolve. Then add 5 g of copper sulphate; dissolve, and add 5 g of potassium iodide. Finally make up to 1.0 dm^3 with distilled water. Each reagent must be fully dissolved before adding the next. The solution should be kept in a dark bottle.)
- Sudan III in ethanolic solution
- Beaker to use as boiling water bath
- Pipettes or syringes
- Test tubes
- Microscope slides and coverslips
- Microscope

IRRITANT
Sodium hydroxide

HIGHLY FLAMMABLE
sudan III in ethanolic solution

Method

1. Estimate the reducing sugar content of the milk samples, using the method described in Practical: Quantitative estimation of sugars.
2. To test for proteins, place 2 cm^3 of the sample to be tested in a test tube and add an equal volume of biuret reagent. A purple-violet colour develops slowly, the intensity of which is proportional to the protein content.
3. To show the presence of fat, add a minute drop of Sudan III solution to a drop of milk on a microscope slide and apply a cover slip. Examine using a microscope; an emulsion of fat droplets should be visible.

Results and discussion

1. Prepare a table to record your observations using each test.
2. Compare the reducing sugar content, protein and fat content of each sample of milk.

2 Nucleic acids, the genetic code and protein synthesis

Nucleic acids

Nucleic acids are macromolecules involved in the storage and transfer of genetic information. They have relative molecular masses ranging from 10^4 to 10^6. They are built up of **mononucleotide** subunits, which join together forming long unbranched chains. The bonds that link the subunits together are sugar–phosphate bonds, sometimes referred to as phosphodiester bonds.
A mononucleotide is made up of:

- a pentose (5–carbon sugar)
- an organic nitrogenous base
- phosphoric acid.

A condensation reaction between the pentose and the base results in the formation of a **nucleoside**, with the removal of a molecule of water. A further condensation reaction between the nucleoside and phosphoric acid produces a mononucleotide. A **phosphoester** linkage is formed and another molecule of water is eliminated (Figure 2.1).

A range of nucleotides can be found in living cells, differing in their pentose and organic bases. Two pentoses, ribose and deoxyribose, are involved. These sugars differ in that deoxyribose ($C_5H_{10}O_4$) has one fewer oxygen atoms than ribose ($C_5H_{10}O_5$) (Figure 2.2). Nucleotides containing ribose, referred to as ribonucleotides, are involved in the formation of ribonucleic acid (RNA) and those containing deoxyribose are deoxyribonucleotides, found in

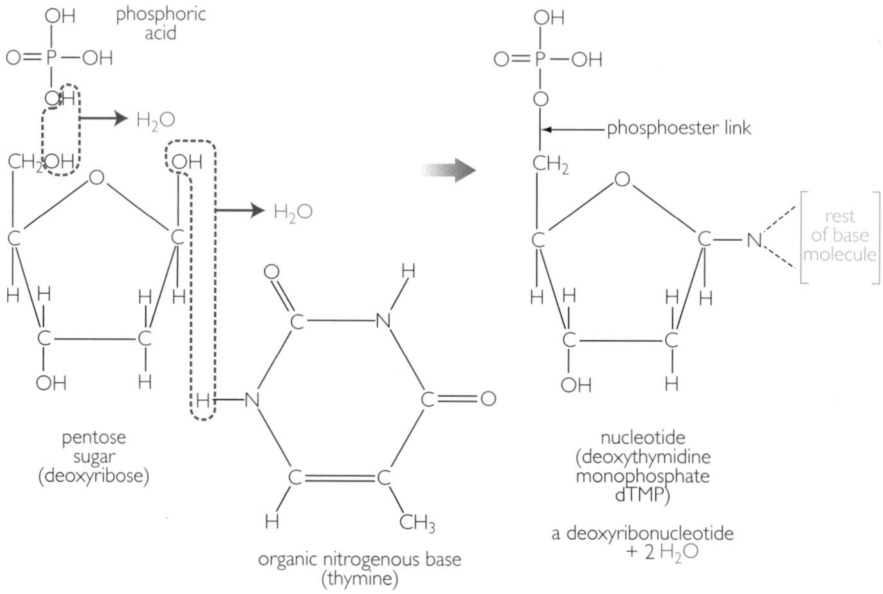

Figure 2.1 Formation of a mononucleotide by condensation of an organic nitrogenous base, a pentose sugar and a molecule of phosphoric acid.

deoxyribonucleic acid (DNA). The organic bases present in nucleotides are either **pyrimidines**, which have a single-ring structure, or **purines** with a double-ring structure. **Cytosine**, **thymine** and **uracil** are pyrimidines and the purines are **adenine** and **guanine**. These bases form links with the carbon–1 of the pentose. Adenine, cytosine and guanine are present in both types of nucleotides. In addition, uracil may be present in ribonucleotides and thymine in deoxyribonucleotides (Figure 2.2).

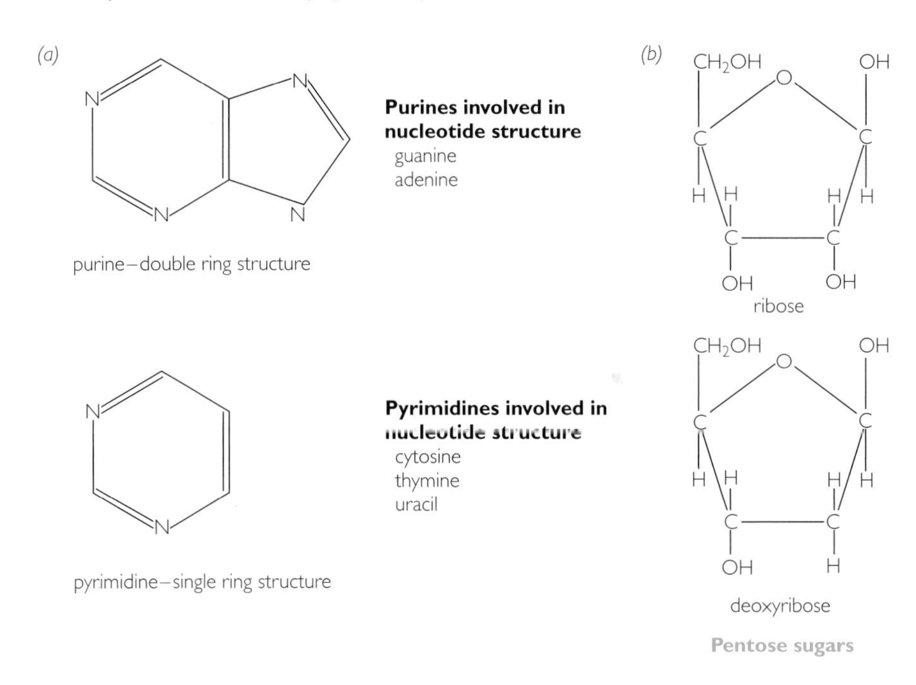

Figure 2.2 Structural formulae of: (a) purine and pyrimidine bases; (b) pentose sugars – ribose and deoxyribose.

A **dinucleotide** is formed when a condensation reaction occurs between the phosphate group of one nucleotide and the pentose of another. More nucleotides can be added, building up a long **polynucleotide** chain. **Phosphodiester** bonds link the nucleotides together (Figure 2.3). Such bonds are covalent and contribute to the stability of the polynucleotide.

Extension Material

Role of nucleotides and related compounds in cells

Apart from their involvement in the formation of the nucleic acids, nucleotides and related compounds have important roles in cells. Adenosine triphosphate (ATP), a ribonucleotide with three phosphate groups attached to the pentose, acts as an energy storage molecule. When the phosphate group at the end is removed by hydrolysis, energy is released, which can be used for metabolic reactions in cells. Molecules of ATP can be built up, using energy from respiration, by the addition of a phosphate group to adenosine diphosphate (ADP). Other derivatives of nucleotides, such as nicotinamide adenine dinucleotide (NAD), nicotinamide adenine dinucleotide phosphate (NADP) and coenzyme A, are involved in metabolic reactions.

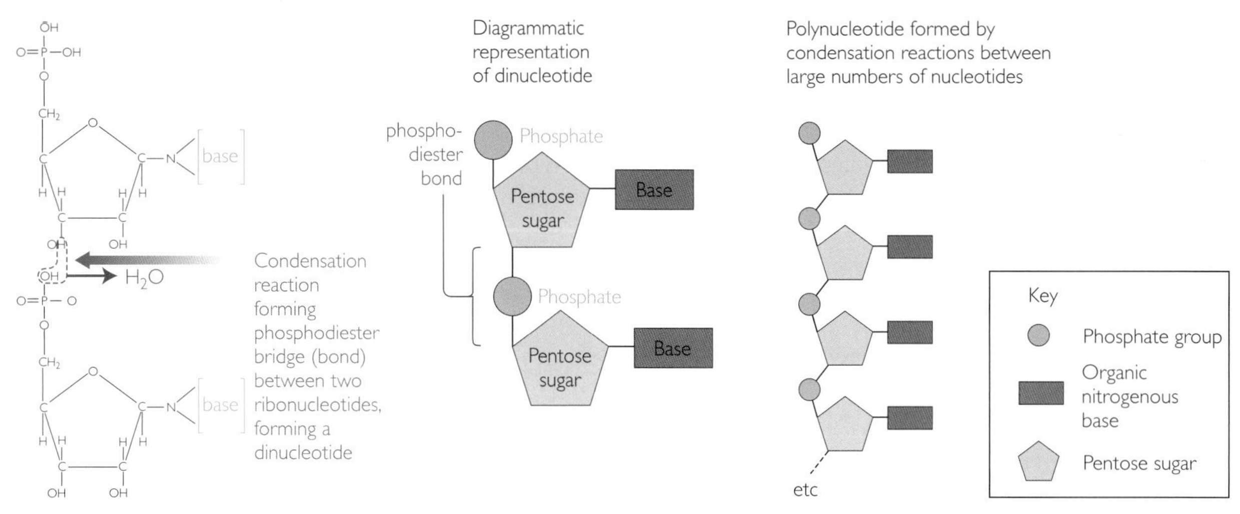

Figure 2.3 Formation of a polynucleotide by repeated condensation of nucleotides.

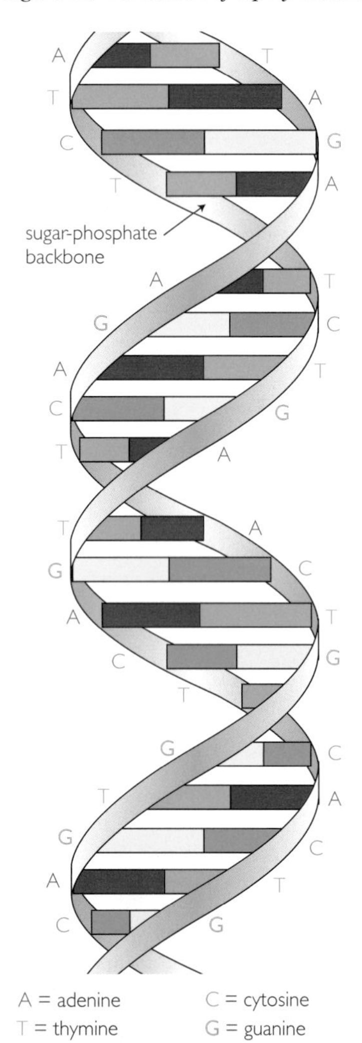

Figure 2.4 Structure of DNA, illustrating the so-called 'double helix' of paired polynucleotide strands running in opposite directions (3′ → 5′ and 5′ → 3′).

The structure of DNA

DNA is a polymer of deoxyribonucleotides. The nucleotides join when covalent, phosphodiester bonds are formed between the carbon–3 of the pentose of one and the phosphate group of the next, giving rise to a sugar-phosphate backbone with the bases projecting outwards. DNA is double stranded, so two polynucleotide chains are involved. Each chain has a 5′ end and a 3′ end. At the 5′ end, the carbon–5 of the pentose residue of the nucleotide is nearest the end, and at the 3′ end it is the carbon–3 of the pentose residue that is closest to the end.

The two chains coil around each other forming a **double helix**, which is held together by hydrogen bonds between 'complementary' bases, called **base pairs**. Pairing occurs only between a purine and a pyrimidine. In DNA, adenine can only pair with thymine and guanine with cytosine. So that the base pairing can occur, the chains are anti-parallel: one chain runs from 5′ to 3′ and the other from 3′ to 5′. It has been determined that the base pairs are 0.34 nm apart and that there are 10 base pairs in one complete turn of the helix (Figure 2.4). Because of the rules of base pairing, the sequence of bases along one of the polynucleotide chains determines the sequence along the other.

Replication of DNA

A molecule capable of acting as the genetic material must have a means of coding for and storing the information. In addition, it must be able to make exact copies of itself. In DNA, the nucleotides can be arranged in any sequence along one of the polynucleotide chains, so there is enormous scope for variation in their arrangement and it is possible for all the necessary information to be coded for. The complementary base pairing means that self-replication is possible and that identical copies can be made.

When Watson and Crick built their model of DNA in 1953, they suggested how replication might occur and subsequent experimental work with bacterial DNA by Kornberg and by Meselson and Stahl provided evidence to support their hypothesis. It was suggested that the two polynucleotide chains of the DNA double helix unwind from one another, due to the disruption of the hydrogen bonds between the base pairs. Each chain then serves as a template for the synthesis of a new complementary polynucleotide chain. It was suggested that the DNA molecule 'unzips' from one end and new nucleotides, present in the nucleus, are assembled in the correct sequence according to the rules of base pairing. This is a complex process and a number of enzymes are required to catalyse the different stages. The enzyme **DNA polymerase** catalyses polymerisation of the polynucleotide chains, but can work only when the chain is built up in the 5′ to 3′ direction; it cannot catalyse polymerisation in the 3′ to 5′ direction. The 3′ to 5′ chain is replicated in short sections, which are then joined together by another enzyme, **DNA ligase**. On completion of this process along the length of the molecule, two identical 'daughter' molecules are formed, each being an exact copy of the original double helix (Figure 2.5).

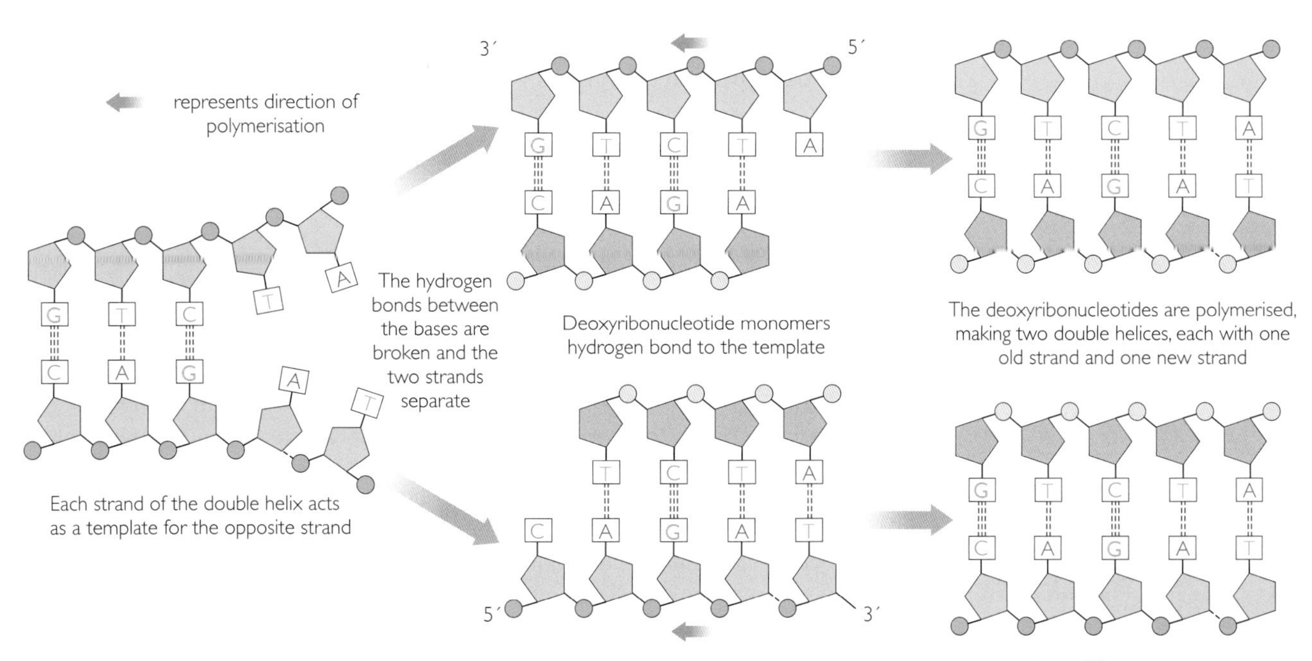

Figure 2.5 Stages in the semi-conservative replication of DNA, in which each polynucleotide strand acts as a template for the synthesis of a new strand.

This form of replication has been referred to as **semi-conservative replication**, because each newly formed double helix contains one of the polynucleotide chains of the original double helix.

As the DNA in each chromosome is so long, replication that starts at one end and proceeds along the length of the molecule would take too much time. Observations made during the replication of *Drosophila* (fruitfly) chromosomes have shown that the double helix opens up at a number of different sites, known as **replication forks**, thus speeding up the process.

The genetic code

A **gene** is a small section of DNA and each chromosome contains thousands of genes. As the only parts of the nucleotides making up the DNA that vary are the bases, then the information must be coded for by the type and sequence of the different nucleotide bases along the polynucleotide chains.

> **DEFINITION**
>
> A **gene** is defined as the sequence of nucleotide pairs along a DNA molecule that codes for an RNA or polypeptide product.

NUCLEIC ACIDS, THE GENETIC CODE AND PROTEIN SYNTHESIS

In the early 1900s, Sir Archibald Garrod suggested that genes exert their effects by means of enzymes. His observations on the inheritance and causes of two human diseases, alkaptonuria and phenylketonuria (PKU), led him to the idea that these conditions are due to the inability of the body to synthesise particular enzymes. In PKU the body cannot convert the amino acid phenylalanine into tyrosine, due to the absence of a single enzyme. This condition can be inherited as a recessive character determined by one gene and suggests a link between genes and enzymes.

Much later, in the 1940s, Beadle and Tatum attempted to find out how much information is contained in a single gene. They designed an experiment to test the hypothesis that one gene codes for one protein.

As a result of their experiments, they proposed the **one gene – one enzyme** hypothesis. As enzymes are proteins, this hypothesis was widened to the **one gene – one protein** hypothesis.

As we realise from studies of their structure, many proteins consist of more than one polypeptide chain. A further modification of the hypothesis to take account of this was suggested by Ingram in 1956. He investigated the structure

Extension Material

Beadle and Tatum's experiments on *Neurospora*

Beadle and Tatum chose to use the bread mould, *Neurospora crassa*, for a number of reasons. It could be easily cultured in the laboratory, crosses could be made between different strains, single spores were not difficult to isolate and the results could be observed within a short time. This mould normally grows on a solution containing a carbon source (sucrose), a nitrogen source (ammonium or nitrate ions), other mineral ions and the vitamin biotin. This solution was referred to as the minimal medium. The mould is able to produce the enzymes necessary to synthesise all the amino acids it requires. They found occasionally that a spore would be produced which was unable to grow and produce a mycelium on the minimal medium, but would grow and reproduce if provided with amino acids (Figure 2.6). This appeared to be a defect in the genetic material (**a mutation**), which could be passed on to the offspring produced from the reproductive structures on the mycelium derived from the mutant spore. The mutation was shown to be due to a single faulty gene. Beadle and Tatum showed that mutant moulds of this type, incapable of growth on the minimal medium, needed only a single amino acid to be added in order to grow and reproduce normally. They concluded that the mutant was unable to synthesise an enzyme needed in the production of that particular amino acid and proposed that one gene contains the information needed for the production of a single enzyme

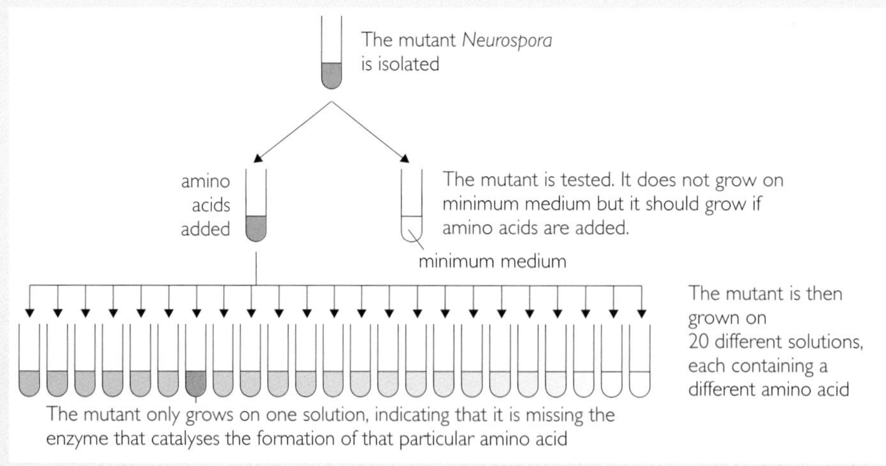

Figure 2.6 Beadle and Tatum's Neurospora *experiment, which showed that a mutation in one gene was sufficient to change the organism's ability to synthesise a particular protein.*

of an abnormal form of haemoglobin found in people suffering from the disorder known as sickle-cell anaemia, in which the red blood cells assume a characteristic sickle shape and do not transport oxygen as well as normal red blood cells. This condition is known to be genetically determined.

Haemoglobin molecules have four polypeptide chains: two identical α-chains and two identical β-chains. These two different polypeptides are coded for by two separate genes. Ingram showed that the difference between the normal haemoglobin and the abnormal form, known as haemoglobin S, was due to one amino acid. In the β-chains, the amino acid valine had been substituted for the amino acid glutamic acid. This evidence provided strong support for a **one gene – one polypeptide** hypothesis.

More recently, the region of DNA that carries information for the production of a polypeptide has been defined as a **cistron**. The hypothesis has been further modified to take account of this new definition and is now referred to as the **one cistron – one polypeptide** hypothesis.

As the nature of a protein is determined by the specific sequence of amino acids in the polypeptide chain, it seemed logical to suggest that the order of the nucleotides in the DNA determines the order in which the amino acids are arranged in a polypeptide. This relationship between the DNA nucleotide bases and amino acids is known as the **genetic code**. As there are 20 common amino acids and only four different nucleotide bases, it was obvious that more than one nucleotide base would have to be involved in coding for an amino acid. A code consisting of two bases for each amino acid would only cater for

Background Information

Crick's experiment on frame shift mutations in viruses

Evidence for this triplet code was provided from the results of some experiments carried out by Francis Crick in 1961. He produced mutations in specific genes on viral nucleic acids by adding or removing one or more nucleotides. This type of mutation is known as a frame shift and changes the nature of the codons, resulting in an alteration to the amino acid sequence (see Figure 2.7). If only one or two nucleotides were involved, the resulting polypeptides were abnormal, with completely different amino acids, and the chains were often shorter than the original ones. If groups of three nucleotides were involved, the polypeptides produced were almost normal as there was little change in the amino acid sequence. If a group of three nucleotides was added, there would be an extra amino acid in the polypeptide chain and if a group of three was removed there would be one amino acid less (Figure 2.7). These experiments also showed that the code was non-overlapping, that is, the bases in a specific triplet do not contribute to any adjacent triplets.

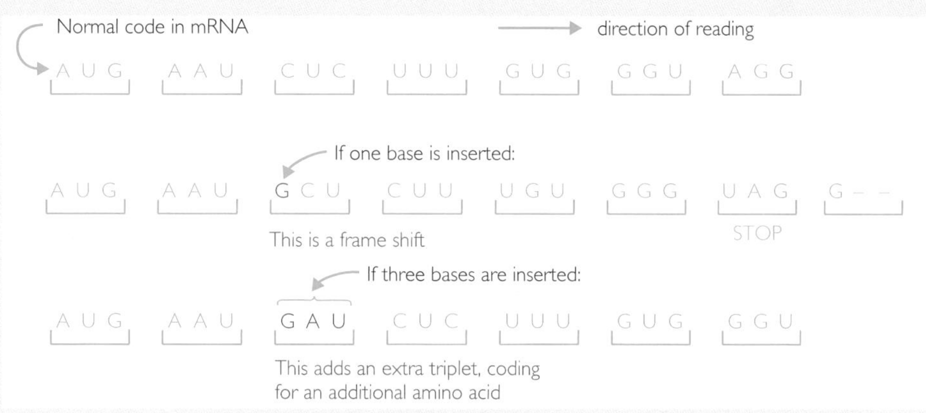

Figure 2.7 Crick's experiment on frame shift mutations in viruses, which showed that a change in one or more mRNA bases in a triplet of three – a codon – resulted in a change to the proteins produced by the host cell. A = adenine; U = uracil; G = guanine; C = cytosine.

16 amino acids, but if the code was three bases (**a triplet**), then there would be 64 possible combinations, more than enough for the 20 amino acids. This is known as the **base triplet hypothesis**, which is now accepted.

DEFINITION

A **codon** is a sequence of three nucleotide bases (a base triplet) on the messenger RNA (mRNA) which codes for a specific amino acid.

Deciphering the code

A series of experiments designed by Nirenberg in the late 1950s resulted in the discovery of the triplet bases that coded for the different amino acids.

As there are 64 possible combinations of bases in the code, some amino acids are coded for by more than one codon, and there are codons that have been found to occur at the ends of polypeptide chains, the so-called chain-terminating, or 'stop', codons. Because the number of codons is greater than the number of amino acids, the code is said to be **degenerate**.

For many years the genetic code was believed to be universal, with the same codons specifiying the same amino acids in all living organisms. Recent studies involving DNA-sequencing techniques have shown that the genetic codes of mitochondrial DNA are slightly different from the standard genetic code. As an example, in mammalian mitochondria, UGA specifies tryptophan rather than 'stop'. The standard genetic code, although very widely used, is not universal.

EXTENSION MATERIAL

Nirenberg's experiment

Nirenberg synthesised molecules of **messenger RNA**, the molecule that carries the information from the nucleus to the ribosomes in the cytoplasm. These synthetic mRNA molecules consisted of the same triplet of bases repeated many times. Twenty tubes were prepared, each containing a different radioactively labelled amino acid, together with ribosomes, enzymes, ATP and other molecules necessary for protein synthesis. Molecules of the specially synthesised mRNA were then added to each of the tubes and, after a period of time, examined to see if a polypeptide had formed. The first synthetic mRNA molecule he used consisted of UUU triplets and resulted in polypeptide formation in the tube containing the amino acid phenylalanine. Having discovered the codon for this amino acid, the codons for all the amino acids were worked out by trying different combinations of bases in the triplets.

DEFINITION

The genetic code is described as **degenerate** because the number of amino acids is less than the number of codons, i.e. a specific amino acid may be coded for by more than one codon.

Protein synthesis

Protein synthesis involves the transfer of the coded information from the nucleus to the cytoplasm (**transcription**) and the conversion of that information into polypeptides on the ribosomes (**translation**). Both these stages involve another nucleic acid, **ribonucleic acid** (**RNA**), which is present in abundance in cells that are synthesising proteins.

Ribonucleic acid is similar in structure to DNA, in that each is a polynucleotide, but it differs in that:

- it is a single-stranded molecule
- it contains the pentose ribose instead of deoxyribose
- the base thymine is replaced by uracil.

There are three different types of RNA present in cells and they all are involved in some way with the synthesis of proteins. They are:

- **messenger RNA** (**mRNA**), which is formed in the nucleus during the process of transcription and which carries the instructions from the DNA to the ribosomes

- **ribosomal RNA (rRNA)**, which is a component of the ribosomes on which the polypeptide chains are built up
- **transfer RNA (tRNA)**, which is present in the cytoplasm and which picks up amino acids and transports them to the ribosomes for assembly into polypeptides.

Transcription

Transcription occurs in the nucleus. The enzyme **RNA-polymerase** becomes attached to the double helix of the DNA in the region of the gene that is being expressed. This usually occurs at a codon for the amino acid methionine, which acts as a 'start' signal. The hydrogen bonds in this region of the double helix are broken and the DNA unwinds. Only one of the strands of the DNA, the **coding strand**, acts as a template and is copied by base pairing of nucleotides. A complementary polynucleotide strand of mRNA is built up from a pool of nucleotides in the nucleus (Figure 2.8). The formation of the strand is catalysed by RNA-polymerase. As it forms, the strand of mRNA detaches from the coding strand of the DNA and, when complete, it leaves the nucleus through a pore in the nuclear envelope. Once in the cytoplasm, it becomes attached to a **ribosome**.

Before the mRNA leaves the nucleus, it is modified by the addition of a guanine molecule to the 5′ end of the polynucleotide chain. This is referred to as a 'cap' and is thought to act as a signal promoting translation, once the mRNA reaches a ribosome. At the 3′ end of the chain, a 'tail' of about 100 adenine nucleotides is added, called poly-A. It is suggested that this tail may act as a signal for the export of the mRNA from the nucleus and it is also considered to offer some protection from enzyme action, as those mRNA strands that lack such tails do not survive for long in the cytoplasm.

> **QUESTION**
>
> How many amino acids are coded for by the MRNA strand in Figure 2.8?

Eukaryotic genes contain regions that do not code for amino acids. These non-coding regions are called **introns** and are copied from the DNA along with

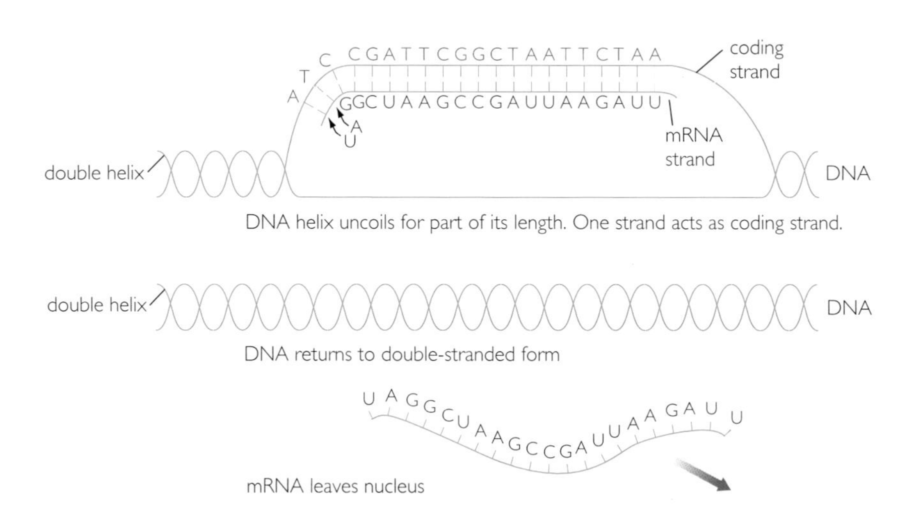

Figure 2.8 Stages in the transcription of the DNA genetic code to produce a mRNA strand that leaves the nucleus for translation in the cytoplasm.

> **DEFINITIONS**
>
> **Anticodon** – a triplet of three exposed nucleotides on a tRNA molecule. These nucleotides are complementary to those of a specific codon on the mRNA.
>
> **Transcription** – the process in which mRNA is synthesised from the information carried on the DNA template. The code is copied.
>
> **Translation** – the process in which the information carried on the mRNA is used in the synthesis of a polypeptide chain. The base sequence of the mRNA determines the sequence of amino acids in the polypeptide.

the coding regions, or **exons**, during transcription. After the 5′ cap and the 3′ tail have been added to the mRNA, and before it leaves the nucleus, the introns are removed by enzyme action (Figure 2.9). The exons are then joined up, so that the mRNA that leaves the nucleus consists of a continuous coding region. The function of the introns is not known.

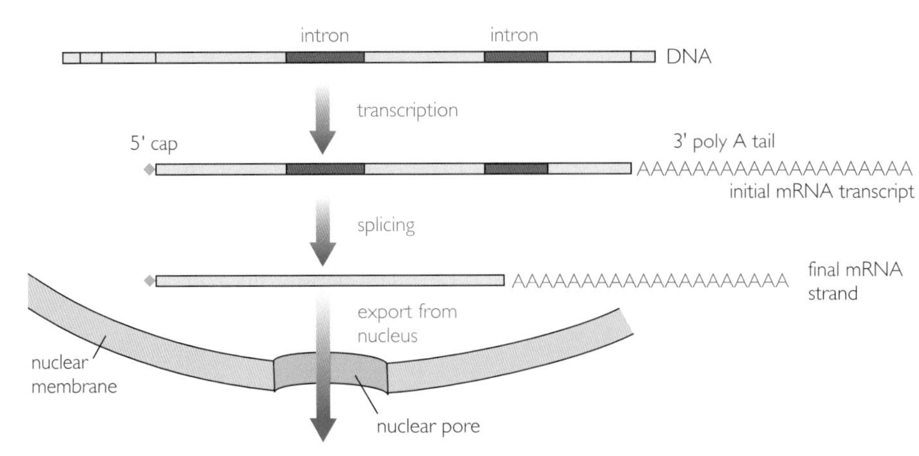

Figure 2.9 Stages in the removal of non-coding lengths of DNA – introns – from the original DNA strand, leading to export of the mature mRNA from the nucleus.

Translation

Translation occurs on the ribosomes in the cytoplasm. The mRNA binds to the small sub-unit of a ribosome and is held in such a way that its codons are exposed. These codons need to be recognised and pair with complementary **anticodons** on the tRNA molecules.

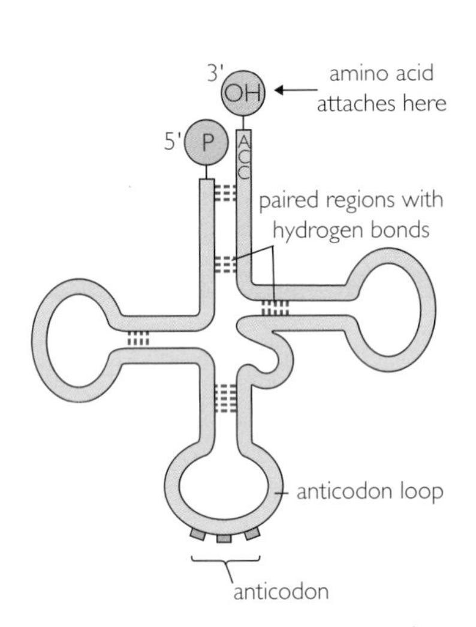

Figure 2.10 Cloverleaf structure of tRNA, showing attachment site for amino acid and anticodon link to mRNA.

All tRNA molecules have the same basic structure. Each consists of a single polynucleotide strand of RNA, about 80 bases long, which is bent back on itself forming a clover-leaf arrangement, held in place by some areas of base pairing. The anticodon consists of a triplet of unpaired bases on one portion of the molecule. The anticodon is complementary to one or more of the codons of mRNA. At the opposite end of the molecule is a site for the attachment of a specific amino acid. The base guanine is always found at the 5′ end of the strand and the 3′ end always has the base sequence CCA.

Before becoming attached to its tRNA molecule, a specific amino acid is activated by ATP. An intermediate is formed, which then joins with the tRNA to form amino acyl-tRNA (Figure 2.10). This reaction is catalysed by an enzyme, amino-acyl tRNA synthetase.

As the ribosome moves along the mRNA strand, two tRNA molecules, with their amino acids, can be held in position at any one time. Their complementary anticodons are held in place opposite the codons on the mRNA by hydrogen bonding. An enzyme, peptidyl transferase, catalyses the formation of a peptide bond between the two amino acids. As the bond is formed, the ribosome moves one triplet further along the mRNA strand (Figure 2.11). Once the amino acids have joined, the tRNA molecules are released and can form a complex with another amino acid of the same type.

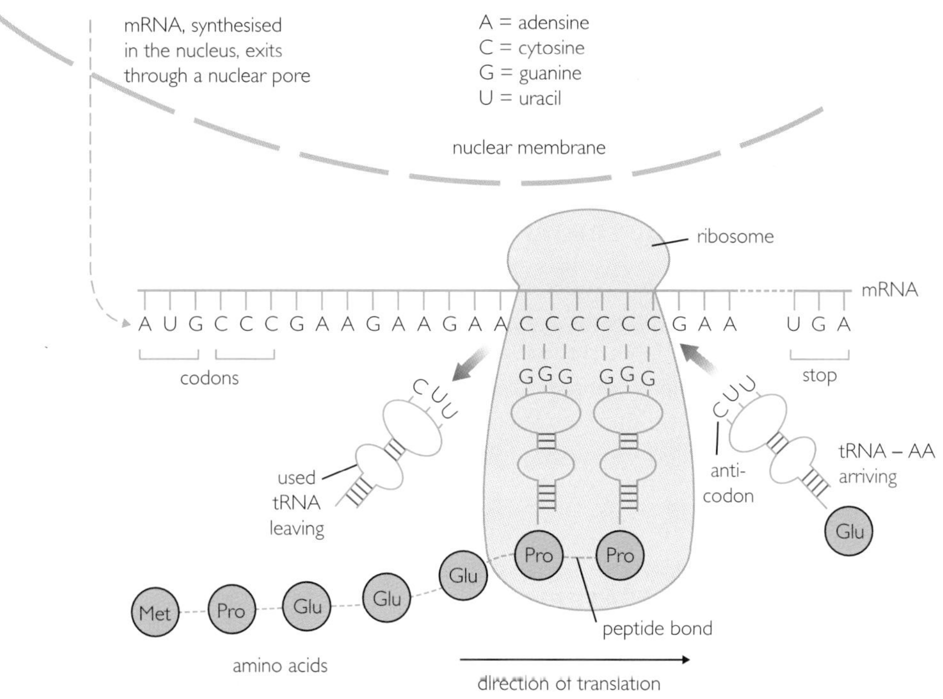

Figure 2.11 Stages in translation of mRNA strand on ribosomes and condensation reactions between amino acids on tRNAs, forming a new polypeptide chain.

Amino acids can be joined to the growing polypeptide chain at the rate of 15 per second. The base pairing between the codons on the mRNA and the anticodons on the tRNA molecules ensures that the transcribed information on the mRNA is exactly translated into the correct sequence of amino acids in the polypeptide chain. Each polypeptide chain usually begins with the amino acid methionine, which is coded TAC on the DNA, giving the codon AUG on the mRNA. The anticodon UAC on the tRNA brings the correct amino acid to the ribosome to start the chain. The sequence is completed when the ribosome reaches a 'stop' codon (UAA, UGA or UAG). On completion, the polypeptide leaves the ribosome. It is not yet a protein, as it has to acquire its secondary structure, coiling into an α-helix or arrangement into a β-pleated sheet, and further folding to form its tertiary structure.

> **QUESTION**
>
> Using the text in this chapter, draw a flow diagram to show how a protein is synthesised. You can check your answer with the diagram on p. 57.

Several ribosomes can attach to an mRNA strand simultaneously, forming **polysomes**, speeding up polypeptide chain formation (Figure 2.12). It has been observed that ribosomes can initiate protein synthesis many times on the same mRNA strand, so that a great deal of protein can be produced from one mRNA molecule.

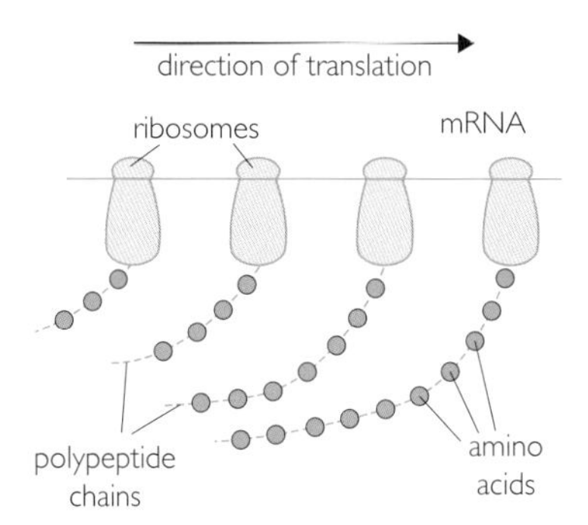

Figure 2.12 A polysome, or polyribosome, is a group of ribosomes translating the same mRNA strand at different points, generating many copies of the polypeptide.

Ribosomes occur both freely in the cytoplasm and attached to the endoplasmic reticulum (ER). Proteins that are destined for secretion are synthesised on the ribosomes attached to the ER and have special signal sequences, which interact with the membrane of the ER. When the polypeptide chains are released from the ribosomes, they pass through the membrane of the ER and are transported in vesicles to the Golgi apparatus, where they may be modified. Vesicles formed from the Golgi apparatus then transport the proteins to the cell surface membrane, where exocytosis occurs and the protein is secreted from the cell (see Chapter 4).

Although reference has been made in this chapter to specific amino acids, their codons and anticodons, it is not necessary for these names to be remembered.

Transfer and ribosomal RNA

Transfer and ribosomal RNA molecules are also coded for by genes, but they are not synthesised in the same way as proteins. They are made directly by transcription from the DNA in the nucleus, in much the same way as mRNA is built up. The genes that code for rRNA, of which there are three different types, are present in multiple copies found in special regions of a chromosome. These regions are called secondary constrictions, or nucleolar organisers.

Extension Material

Structural and regulatory genes

There are two main types of genes that code for polypeptides:

- structural genes, which code for functional proteins such as enzymes, hormones, antibodies, storage proteins and fibres
- regulatory genes, which control the activities of other genes.

All the body cells of an organism contain the same genetic information. They have the same number of chromosomes with the same number of genes, so it is logical to suggest that there is some mechanism by which the activity of genes is controlled. In the early 1950s, Jacob and Monod designed a series of experiments to investigate how enzyme synthesis was controlled in the gut bacterium *Escherichia coli* (Figure 2.13). It was known that some enzymes were produced all the time, but others were only synthesised if a particular compound was present. Jacob and Monod found that *E. coli* would grow well if supplied with glucose, but if supplied with lactose growth would stop for a short time, after which it would grow rapidly again. Their investigations showed that in order to break down the lactose, the bacterium needed to synthesise other enzymes that were not normally present. In 1961 they suggested that the structural genes coding for these enzymes were present on the DNA, but their transcription and translation were suppressed by a regulator gene. The regulator gene codes for a so-called repressor molecule, which binds to part of the DNA, preventing the intitiation of transcription of the codons for the enzymes that enable the bacterium to take up and metabolise lactose. When lactose is present, it binds to the repressor molecule and inactivates it. As soon as the lactose is removed or used up, synthesis of the enzymes ceases within a short time. Jacob and Monod mapped the location of these genes on the bacterial DNA and found that they were side by side and close to two other regions, called the promoter and operator regions, which are also involved in the control mechanism. The mechanism is shown in Figure 2.14.

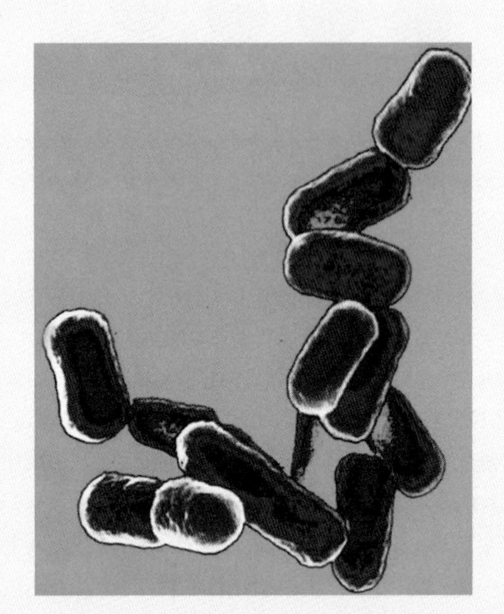

Figure 2.13 Electronmicrograph of the gut bacterium Escherichia coli, *used by Jacob and Monod in their studies of structural and regulatory genes.*

Jacob and Monod called the length of DNA which consists of the structural genes for the lactose breakdown, together with the promoter and operator regions, an operon. Several other operons have been described for bacteria, but no such mechanisms have been found in eukaryotes. In bacteria, transcription and translation are more closely linked than they are in a eukaryote, where transcription occurs in the nucleus and translation in the cytoplasm.

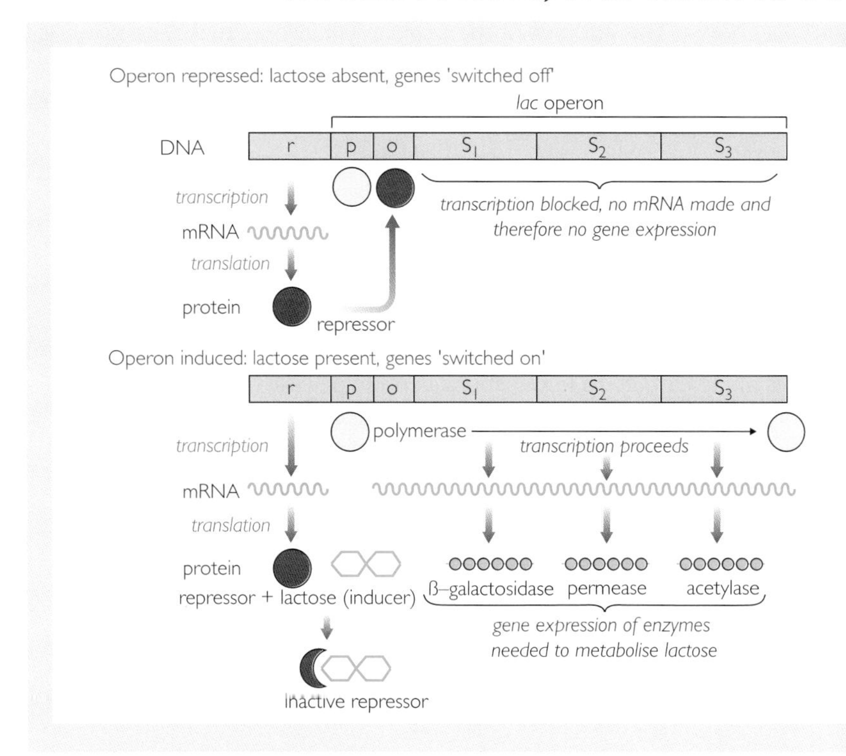

Figure 2.14 Jacob and Monod proposed that, in the absence of lactose, a repressor protein blocks transcription of the genes that code for the production of enzymes needed to metabolise lactose. If lactose is present it binds to the repressor and the genes that lead to enzyme production are 'switched on'.

The Human Genome Project

The genome consists of all the genes in the chromosomes of an individual. It includes the positions and sequences of the genes on the chromosomes, together with the base sequences within the genes.

The Human Genome Project was started in 1990 with the aim of identifying all the genes in human DNA, mapping the positions of the genes on the chromosomes, and determining the sequences of the base pairs along the DNA. In addition, the project set out to develop techniques for the analysis and use of the genome, and to store the information on a database for easy accessibility. It was also necessary to consider the ethical, legal and social implications and the use which could be made of this information.

It was found that there were about 3 billion (10^9) base pairs, but the actual numbers of genes located and mapped was only 18 406. Only a very small proportion (about 1 per cent) of the genome consists of genes; the remainder of the DNA appears to be redundant and non-coding, existing between genes and as introns within genes.

It has been found that many genes present in the human genome are also present as identical genes in other organisms. The differences between ethnic groups and between individuals are very small. Perhaps the most significant discoveries were the so-called 'errant genes' which cause diseases, for example, the gene for Huntington's chorea was located on chromosome 4, while genes relating to Alzheimer's disease were discovered on chromosomes 1, 14 and 21.

The medical implications of the findings of this project are that the detection of errant genes could make the diagnosis of certain diseases easier and could make it possible accurately to predict the onset of diseases such as

Huntington's chorea. The findings of the project will also increase the potential to treat genetic and acquired diseases by gene therapy, whereby defective genes are either replaced by normal genes or supplemented with other genes to prevent a disease taking its course.

There are many ethical, legal and social issues arising from this project. Some of these issues are general while others directly affect individuals. Examples are listed below.

- The ownership and use of personal genetic information, and who should have access to it, are a matter for concern.
- While it is an advantage to know about genetic diseases, so that there is the possibility of treatment, it may be a disadvantage if this knowledge is used in the wrong way, or without the permission of the individual concerned.
- Some genetic tests have been developed and are known to be reliable, but there must be questions about some of the newer techniques, for example, how much can their reliability or interpretation be trusted?

In the future, it could become as common to know one's genetic profile as knowing one's blood group is today.

Enzymes

Enzymes as organic catalysts

Enzymes are substances that act as **catalysts**, in other words they increase the **rate** of chemical reactions. Consider the following general reaction between two substances, A and B, which react together to form a product, substance C:

$$A + B \rightarrow C$$

In biological systems, this reaction might occur very slowly, or not at all, in the absence of an enzyme. Enzymes will greatly increase the rate of formation of the product. Enzymes can increase the rate of reactions by a factor of at least one million. One of the fastest enzymes known is carbonic anhydrase, which catalyses the following reaction:

$$CO_2 + H_2O \rightleftharpoons H_2CO_3$$

Carbonic anhydrase reacts with 10^5 molecules of carbon dioxide per second.

Most enzymes are large globular protein molecules, with complex three-dimensional shapes (see Chapter 1, page 17), but recently it has become clear that substances other than proteins have catalytic properties.

Unlike inorganic chemical catalysts, such as iron and nickel, enzymes are specific. This means that each enzyme normally only catalyses one reaction. The substance with which the enzyme combines is known as the **substrate**. This combines with the enzyme at a particular place on the enzyme's surface called the **active site**. Enzyme molecules are usually very much larger than their substrates, and the active site is only a relatively small part of the enzyme, consisting of only 3 to 12 amino acid residues. The rest of the enzyme molecule is involved in maintaining the shape of the active site. The precise shape of the active site is important because it is complementary to the shape of the substrate molecule, which fits into the active site by what is often known as the 'lock-and-key mechanism'. However, it is now clear that the shape of the active site of some enzymes changes when the substrate molecule attaches to it. This process is referred to as **induced fit**. The substrate molecule joins with the active site of the enzyme to form an enzyme–substrate complex (Figure 3.1).

Energy and chemical reactions

When chemical reactions occur, energy changes result from the changes in structure of the reactants, breaking and reforming chemical bonds. These energy changes are in the form of heat. Some reactions produce heat, others absorb heat. A reaction such as combustion, which produces heat, is known as **exothermic**, but if heat is taken in, it is called an **endothermic** reaction. Most reactions that occur spontaneously are exothermic. In living organisms, energy produced in a reaction can take various forms, such as heat, light or chemical energy. Heat and light will be dissipated, and therefore will not be available to

Extension Material

Discovery of catalytic RNA

In 1983, Thomas Cech at University of Colorado discovered a type of RNA with enzymic properties in the ciliate protozoan *Tetrahymena*. This catalytic RNA was termed **ribozyme**. It is involved in the removal of part of a large RNA molecule and the subsequent joining together of the remaining parts to form a functional RNA molecule. Similar activity has since been found in the cells of some bacteria and fungi.

Question

Where in the body would you expect to find carbonic anhydrase?

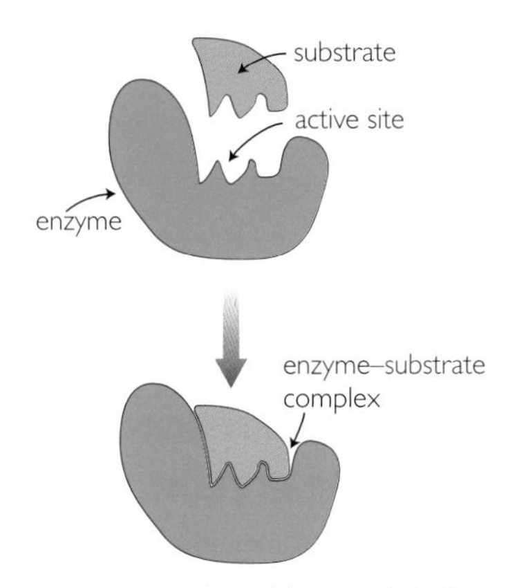

Figure 3.1 Lock-and-key model of enzyme action, in which a substrate molecule and an enzyme molecule interact at a specific active site on the enzyme, forming a temporary enzyme–substrate complex.

the organism for other processes. Other forms of energy can be used by the organism to drive other processes. These forms of energy are known as 'free energy'. Change in free energy is given the symbol ΔG. A reaction only occurs spontaneously if ΔG is negative. In this case the reaction is termed **exergonic**. If ΔG is positive, an input of free energy will be needed to drive the reaction, which is said to be **endergonic**. In biological systems, endergonic reactions include the synthesis of macromolecules, such as proteins, and these reactions are linked to exergonic reactions, which provide the free energy required.

In living cells, most chemical reactions require an input of energy before the molecules react together. This is referred to as the **activation energy.** Imagine a boulder near the top of a steep hill (Figure 3.2). Before the boulder can roll down the other side of the hill, you would have to push it to the top, in other words put some energy in, before it will roll down.

This can be likened to the progress of a chemical reaction: the boulder near the top represents the energy level of the reactants and the energy you have to put in to push the boulder to the very top represents the free energy of activation. Once the reactants reach this point, the reaction will proceed spontaneously, as the boulder rolls down and the products are formed.

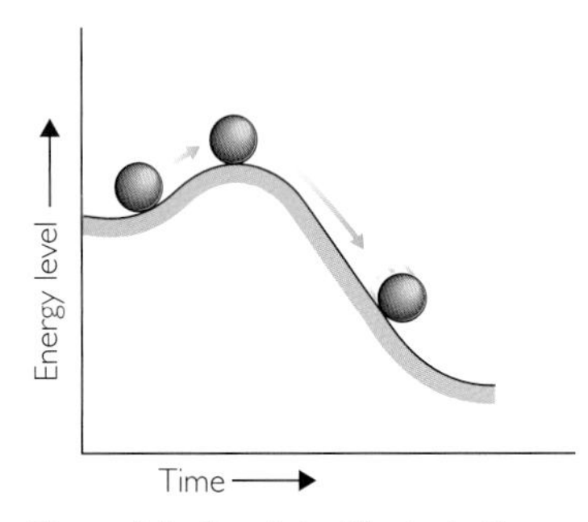

Figure 3.2 Graph to illustrate the boulder analogy of a reaction in which activation energy needs to be added to a system to initiate the reaction.

Enzymes increase the rates of reactions by reducing the free energy of activation, so that the barrier to a reaction occurring is lower in the presence of an enzyme. The combination of enzyme and substrate creates a new energy profile, for the reaction, with a lower free energy of activation (Figure 3.3).

Once the products have been formed, they leave the active site of the enzyme, which is left free to combine with a new substrate molecule. Enzymes, like chemical catalysts, are not used up in the reaction they catalyse so they can be used over and over again. The overall reaction between an enzyme and its substrate can be represented by the following equation:

ENZYME + SUBSTRATE → ENZYME–SUBSTRATE complex → ENZYME + PRODUCTS

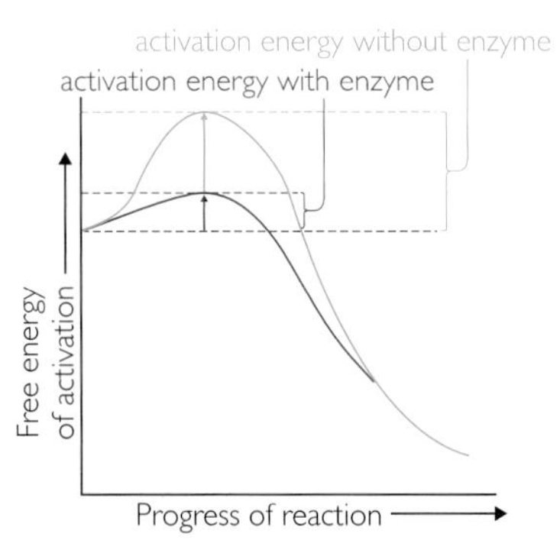

Figure 3.3 Graph to illustrate how addition of an enzyme lowers the amount of activation energy needed to initiate the reaction.

Factors affecting enzyme activity

Enzymes, being proteins, are sensitive to changes in their environment. Changes in temperature and pH can cause changes in the bonding and shape of the enzyme molecule and will therefore affect its activity. Changes in the concentration of both the enzyme and its substrate will also affect the rate of an enzyme-catalysed reaction.

Temperature

Temperature has a complex effect on enzyme activity (see *Practical: The effect of temperature on the activity of trypsin*). On one hand, a rise in temperature increases the kinetic energy of enzyme and substrate molecules, and therefore will tend to increase the rate of a chemical reaction. As the temperature increases, it increases the chance of enzyme–substrate complexes forming, because the substrate molecules are more likely to 'bump into' the enzyme's active site.

Generally, with every 10 °C increase in temperature, the rate of reaction doubles. This is referred to as the **Q_{10} effect**, which is expressed as the rate at (t + 10 °C) ÷ rate at t °C, where t = temperature. However, increases in temperature also affect the stability of the enzyme molecule. Remember that the precise shape of the active site is essential for catalytic activity so, if shape changes because of a change in the bonds, the enzyme will be unable to combine with its substrate. The overall rate of activity will depend on a balance between these two factors and the enzyme will have an apparent optimum temperature at which it functions most rapidly (Figure 3.4).

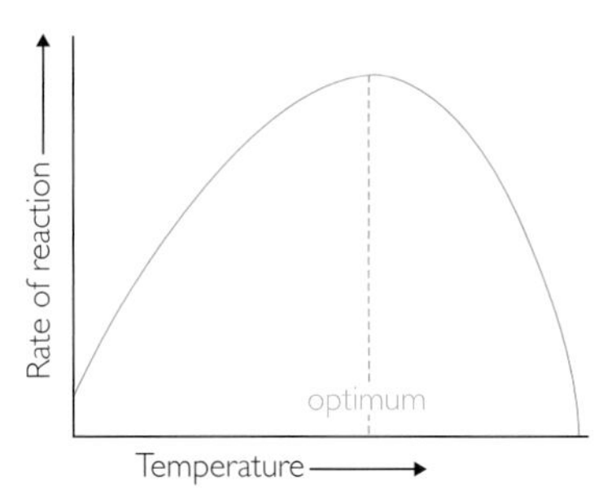

Figure 3.4 Graph to illustrate the effect of temperature on enzyme activity, showing that enzymes have an optimum temperature at which reaction rate is at its maximum.

At temperatures above this optimum the enzyme rapidly loses its activity and becomes progressively **denatured**, that is, it is unable to combine with the substrate and therefore has lost its catalytic properties. It should be noted that denaturation is time-dependent, so exposure to a high temperature for a brief period of time will have less effect on the enzyme than prolonged exposure.

The optimum temperature for enzymes is variable. For example, an enzyme known as Taq polymerase, a DNA polymerase extracted from *Thermus aquaticus* (a bacterium that lives in hot springs), is stable up to 95 °C. Many enzymes, however, function most efficiently at about 40 °C.

> **DEFINITION**
>
> **Denaturation** occurs when the shape of the enzyme molecule. If this changes the shape of the active site, so that the substrate is less likely to bond, the rate of reaction decreases.

pH

Most enzymes have a characteristic pH at which they function most efficiently (see *Practical: The effect of pH on the activity of catalase*). This is known as the optimum pH. Changes in pH affect the ionisation of side groups in the enzyme's amino acid residues. Changes in the bonding within the protein molecule can occur. This in turn affects the overall shape of the enzyme molecule and affects the efficiency of formation of enzyme–substrate complexes. At extremes of pH the enzyme molecule may become denatured.

As with optimum temperatures, enzymes have varying optimum pH values. Although many enzymes have optima at pH values of around 7 (neutral), there are some that function best at extreme values. As an example, pepsins, which are protein-digesting enzymes found in the stomach, have unusually low pH optima, in the range 1.5 to 3.5. At the other extreme, arginase, an enzyme involved in the synthesis of urea in the liver, has an optimum pH of 10. The optimum pH for an enzyme is not necessarily the same as the pH of its normal surroundings. This is one way in which the intracellular environment can control enzyme activity. A typical graph of rate of reaction plotted against pH, produced experimentally by determining the activity of an enzyme at a range of specific pH values, shows a bell-shaped curve. You will notice that, unlike the curve for enzyme activity plotted against temperature, this is a symmetrical shape (Figure 3.5).

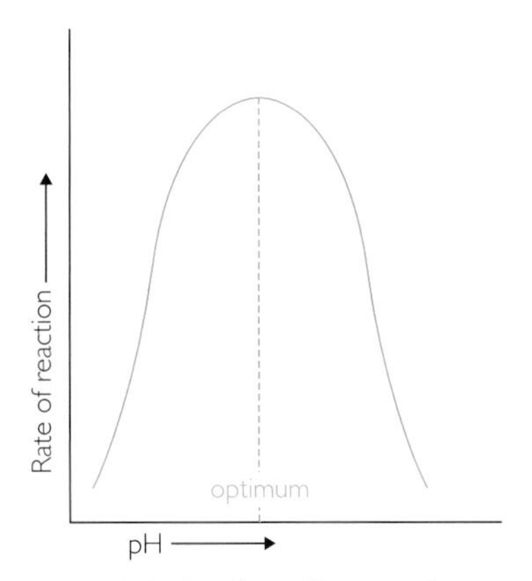

Figure 3.5 Graph to illustrate the effect of pH on enzyme activity, showing that enzymes have an optimum pH at which reaction rate is at its maximum.

Concentration

Both enzyme and substrate concentration affect the rate of reaction (see *Practical: The effect of enzyme concentration on the activity of amylase*). Clearly, as the concentration of enzyme increases, the number of active sites also increases. Provided that there is an excess of substrate molecules, the rate of reaction increases in proportion to the concentration of enzyme. A graph of rate of reaction plotted against enzyme concentration is a straight line (Figure 3.6).

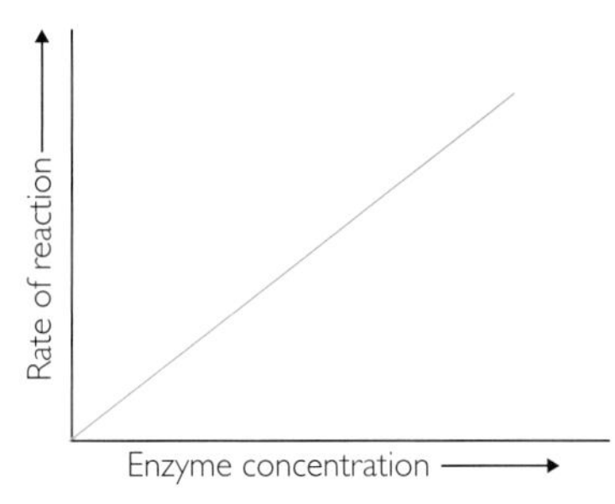

Figure 3.6 Graph to illustrate the effect of enzyme concentration on the rate of a reaction, showing that addition of enzyme increases reaction rate linearly.

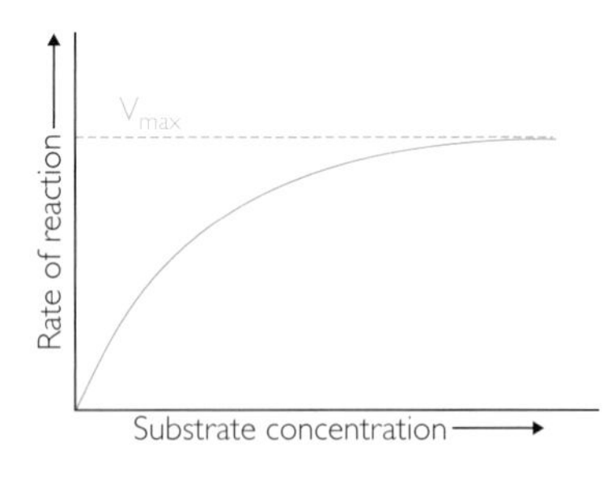

Figure 3.7 Graph to illustrate the effect of substrate concentration on the rate of a reaction, showing that addition of substrate increases reaction rate until the enzyme is saturated, at which point the maximum rate of the reaction (V_{max}) is reached.

Consider what will happen if the substrate concentration increases, but the concentration of enzyme molecules remains the same. Since the rate of reaction depends on the rate of formation of enzyme–substrate complexes, the rate of reaction will increase as the substrate concentration increases, but only until all enzyme molecules are being used. After this point, no matter how much more substrate is added, the enzymes are working as fast as they can, so the rate reaches a maximum velocity and remains constant. This rate is referred to as V_{max}. Figure 3.7 shows the effect of substrate concentration on the rate of reaction.

Inhibitors

Enzyme inhibitors are substances that reduce the rate at which reactions catalysed by enzymes take place. They do this by affecting the active site and so affect the ability of the enzyme to form enzyme–substrate complexes. Broadly, enzyme inhibitors are divided into two groups, known as:

- active site-directed inhibitors, and
- non-active site-directed inhibitors.

Active site-directed inhibitors have a shape which is similar to the shape of the substrate for the particular enzyme, and so are able to combine with the active site (see Figure 3.8a). This blocks the active site, and so prevents the substrate from binding. Some active site-directed inhibitors are able to combine permanently with the active site, so that the enzyme is completely inactivated. An example of such an inhibitor is di-isopropylphosphofluoridate (DIFP), which combines with the active site of the enzyme acetylcholinesterase.

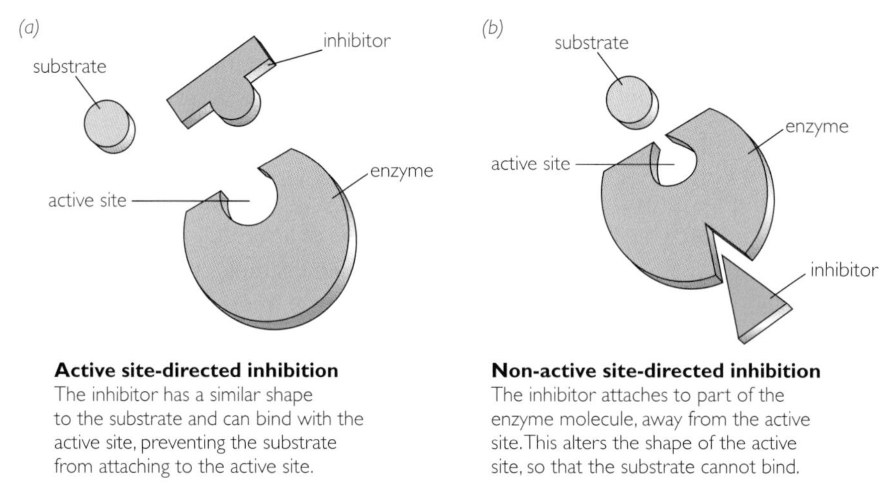

Figure 3.8 Site-directed inhibition: (a) active (b) non-active.

However, some active site-directed inhibitors bind temporarily to the active site and so are able to leave. This means that, when the inhibitor is occupying the active site, the rate of reaction is reduced but, when the inhibitor leaves the active site, the substrate is able to combine. The rate of reaction is therefore reduced, rather than completely stopped. These types of inhibitors are sometime also referred to as competitive inhibitors because they compete with the substrate to occupy the active site. An example of such an inhibitor is malonate, which has a similar shape to the substrate for the enzyme succinate dehydrogenase. Malonate competes for the active site with the correct substrate, succinate, and so slows down the reaction between succinate dehydrogenase and succinate. The overall effect of such an inhibitor depends on the relative concentrations of inhibitor and substrate – the higher the concentration of substrate compared with the inhibitor, the greater the chances of formation of enzyme–substrate complexes.

Non-active site-directed inhibitors are substances that attach to part of the enzyme molecule (known as the allosteric site) away from the active site (see Figure 3.8b). This can change the shape of the enzyme and alter the shape of the active site so that it is unable to form an enzyme–substrate complex. As with active site-directed inhibitors, non-active site-directed inhibitors may also attach to the enzyme permanently or reversibly. An example of a non-active site directed inhibitor is adenosine triphosphate (ATP), which attaches to the enzyme phosphofructokinase, one of the enzymes involved in cell respiration. ATP is one of the products of cell respiration, so when ATP levels are high, enzyme inhibition by ATP decreases the production of ATP. Conversely, when ATP levels fall, enzyme inhibition is reduced and so the production of ATP increases. This is an example of control of an enzyme pathway.

Extension Material

Enzymes and metabolic pathways

The sum total of all the reactions occurring in cells is referred to as **metabolism**. Metabolism consists of hundreds of linked chemical reactions, which make up particular metabolic pathways, such as the breakdown of glucose to produce carbon dioxide and water. These reactions usually occur in a series of small steps, rather than as just one reaction. A metabolic pathway can be illustrated by the following simple diagram, where each letter represents one substance:

$$A + B \rightarrow C \rightarrow D + E$$

Each single reaction in this series will be catalysed by a specific enzyme, so we could add these to the pathway:

$$A + B \xrightarrow{\text{enzyme X}} C \xrightarrow{\text{enzyme Y}} D + E$$

If A and B represent the reactants, and D and E the final products, then substance C is referred to as an intermediate, which is converted by enzyme Y to the final products. To illustrate a metabolic pathway, here is the first part of the process known as glycolysis. Glycolysis is a series of reactions involving the breakdown of glucose, via a series of intermediates, to form a substance called pyruvate.

$$\text{glucose} + \text{ATP} \xrightarrow{\textbf{hexokinase}} \text{glucose 6-phosphate} + \text{ADP} + H^+$$

$\downarrow$ **phosphoglucose isomerase**

fructose 6-phosphate

Notice that each of these two reactions is catalysed by a specific enzyme; in one it is hexokinase and in the other it is phosphoglucose isomerase. (This is developed further in *Respiration and Coordination*, Chapter 1).

Extension Material

Classification of enzymes

There are very many different enzymes (over 2000 have been named and studied in detail), and they are grouped according to the type of reaction they catalyse. There are six major groups, as shown below.

Table 3.1 *Enzyme classification according to reaction type*

Group	Type of reaction catalysed	Example of enzyme
oxidoreductases	removal or addition of hydrogen atoms (oxidation or reduction)	succinate dehydrogenase
transferases	transfer of a group from one compound to another	aminotransferase
hydrolases	hydrolysis	sucrase
lyases	elimination of a group to form a double bond	citrate lyase
isomerases	intramolecular rearrangements	phosphoglucose isomerase
ligases	formation of a bond, coupled with ATP hydrolysis	glycogen synthetase

Commercial uses of enzymes

Traditional fermentations exploit microorganisms and their enzymes to modify foods. These fermentations continue to be very important on an industrial scale, but a newer industry of **enzyme technology**, using enzymes harvested from microorganisms, has grown up alongside them and now has many commercial applications. Enzymes produced from fungi and bacteria can be isolated from the growth media and purified as necessary. When used in an industrial process, the enzyme is often **immobilised** which improves its stability, allowing the enzyme to be re-used and the products to be more easily separated. Microbial enzymes are now used in a wide range of industrial processes, including the production of paper, textiles, leather, fruit juices and biological detergents. They are also being applied in medicine, for diagnosis and treatment.

Pectinases

Pectinases degrade pectins (polysaccharides found within plant cell wall structure) to shorter molecules of **galacturonic acid** (Figure 3.9). This can be broken down further to sugars and other compounds. Pectin itself is able to form jellies, which is useful in some products (such as the setting of jam), but undesirable in fruit juices and other liquids. Pectinases are obtained commercially from fungi, particularly species of *Aspergillus* and *Penicillium*. Bacterial and other fungal pectinases are significant in soft rot of fruits and vegetables, contributing to spoilage and decay. Probably the biggest industrial use of pectinases is in the extraction and clarification of fruit juices. Pectinases are added to crushed fruit, such as apples and grapes, to increase the yield of juice extracted and improve

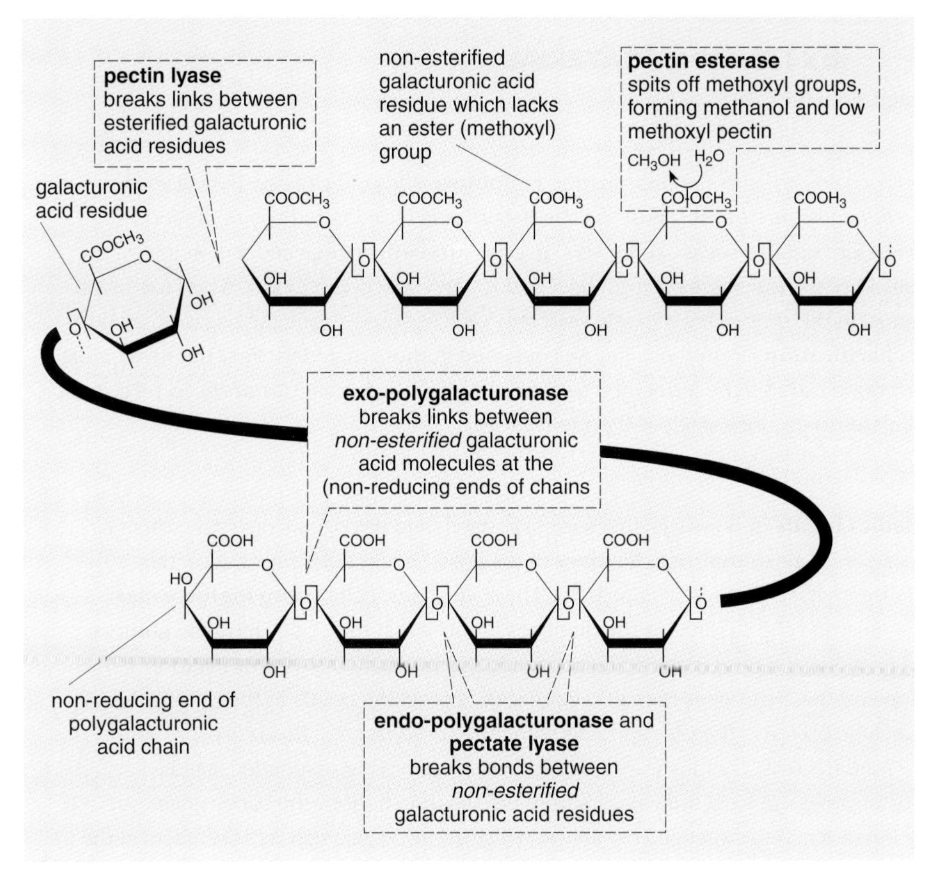

Figure 3.9 Activities of pectinases – a group of enzymes which break down complex pectin molecules to simpler residues. Pectinases are important in the fruit juice industry.

(a)

(b)

Figure 3.10 Small-scale versions of commercial methods used to extract juice from fruit, such as apples: (a) crusher, (b) press. On a commercial scale, addition of enzymes, such as pectinases, can noticeably increase the volume of juice extracted.

colour derived from fruit skins. Wines, vinegar and other liquids which contain suspended pectic material can be clarified with pectinases. They act by removing some of the pectin around charged protein particles, which then clump together and settle out of the liquid. Pectinases can also be used to prevent jelling when fruit juices are concentrated. (See also *Practical: The use of pectinase in the production of fruit juice*.)

Proteases

Proteases are known as proteinases and peptidases. They hydrolyse **peptide bonds** in proteins and peptides, acting either within the peptide chain or removing amino acid residues sequentially from one or other end of the chain. Fungal sources include species of *Aspergillus*, *Mucor* and *Rhizopus*. *Bacillus* spp. provide a source of bacterial enzymes. Proteases account for about 50 per cent of the commercially-used microbial enzymes.

Some fungal enzymes are used in cheese-making as a substitute for rennet to help clot milk. Other uses for microbial proteases include clarification of fruit juices and beer by removing the protein haze, thinning egg white so that it can be filtered before drying, tenderisation of meat, digestion of fish livers to allow better extraction of fish oil and modification of proteins in flour used for bread-making.

EXTENSION MATERIAL

Other carbohydrates and commercial enzymes

Cellulases

Cellulases break down cellulose to shorter chains, then to the disaccharide **cellobiose** and to β-glucose. Fungal sources include species of *Aspergillus*, *Trichoderma* and *Penicillium*. These cellulases currently have limited use in the food industry, but can be used to produce more fermentable sugars in brewers' mashes, to clarify orange and lemon juices, to improve the release of colours from fruit skins, to clear the haze from beer and to tenderise green beans. When used with lignases, cellulases may have great potential in the processing of waste materials such as straw, sugarcane bagasse, sawdust and newspaper, to produce sugars (**saccharification**) from the cellulose contained in these materials. First the wood must be treated to remove lignin. Sugars from wood can then be fermented to alcohol (ethanol). A yeast (*Candida* sp.) has been grown on wood pulp hydrolysed by cellulases to produce single-cell protein (SCP).

Amylases

Amylases hydrolyse **glycosidic** (**glucosidic**) **bonds** in polysaccharides such as starch and glycogen, converting them to **dextrins** (shorter length chains of glucose units) or to **maltose**. Enzymes in the group act in different ways on the α-1,4 links, and on the α-1,6 links within the polysaccharides (Figure 3.10). These amylases include **amyloglucosidase** (glucoamylase) which hydrolyses the 1,4 links and the 1,6 links at the branches in the starch molecule. Terminal **glucose** units are removed from the end of the chain one at a time, rather than giving the intermediate dextrins or maltose. **Pullulanase**, also known as debranching enzyme, hydrolyses the α-1,6 links at the branching points in the polysaccharide. Commercial sources of these enzymes include bacteria (*Bacillus* spp.) and fungi (*Aspergillus* spp., *Rhizopus* spp. and *Streptomyces* spp.). Fungal amylases are used to clarify fruit juices, wines and beer by removing suspended starch.
In bread-making and brewing, addition of amylases can yield more sugars from the starch in flour or the barley grains. An important commercial use is the conversion of starch to sweet glucose syrups which are used generally as sweeteners in the food industry as well as in bread-making and brewing. Altering the balance between amyloglucosidase and the fungal α-amylase can produce different proportions of glucose and maltose. A higher proportion of glucose is useful for fermentation whereas higher maltose is more useful in preparation of jam and confectionery. Further conversion of glucose, using the enzyme glucose isomerase, yields fructose which is sweeter than both sucrose and glucose. High-fructose corn syrups (HFCS), derived from hydrolysing corn (maize) starch, have become a major source of sweeteners in foods and drinks in the USA.

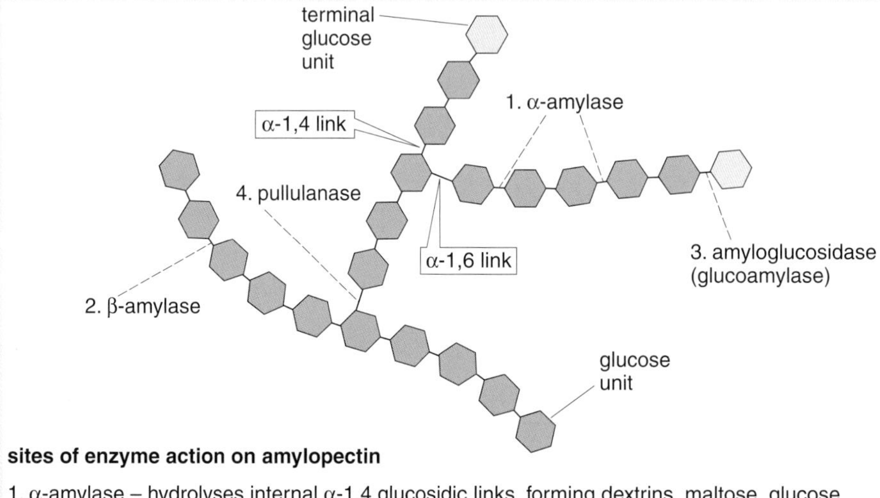

Figure 3.11 Activities of amylases, a group of enzymes which act on glucosidic bonds and break down complex carbohydrates, such as starch and glycogen, to simpler residues.

Lactase

Lactase (**β-galactosidase**) breaks down the disaccharide lactose, the sugar in milk, to **galactose** and **glucose**. Commercially produced microbial lactases are obtained from *Aspergillus* spp. and the yeast *Kluyveromyces* sp. An important application of lactase is to hydrolyse lactose in milk to make it suitable for people who are intolerant of lactose. Such people are unable, as adults, to digest lactose, so the undigested lactose is likely to be fermented by bacteria in the large intestine, resulting in nausea, abdominal pain and diarrhoea. Glucose and galactose both taste sweeter than lactose, so lactase is used to increase sweetness in products such as ice cream and to produce a sweet syrup from whey, which might otherwise be discarded as a waste product from the cheese industry. In ice-cream manufacture, use of lactase also removes lactose which crystallises at low temperatures and would contribute to a 'sandy' texture.

Table 3.2 *Examples of some commercial enzymes, their sources, properties and uses*

Enzyme group	Enzyme trade name	Organism source	Optimum temperature and pH	Industrial applications
cellulase	Celluclast® 1.5L	fungus *Trichoderma reesei*	65 °C pH 5.0	• brewing – reduces wort viscosity (added at mashing in stage) • breakdown of cellulosic material for production of fermentable sugar
pectinases	Pectinex™ Ultra SP-L	fungus *Aspergillus niger*	35 °C pH 5.5	• disintegrates cell walls • for treatment of fruit mashes (especially apples and pears) – increased press capacity and higher juice yields
α-amylase	Termamyl®	bacterium *Bacillus licheniformis*	95 °C pH 7.0	• useful because of its heat stability • alcohol industry – thinning of starch in distilling mashes • brewing – liquefaction, simpler cooking programme • sugar industry – breakdown of starch in cane juice, so less starch in raw sugar
α-amylase glucoamylase	Amylo-glucosidase AMG	fungus *Aspergillus niger*	60 °C to 75 °C pH 4.0	• sweeteners – production of glucose syrup • wine and beer – removal of starch haze • bread – improved crust colour • low carbohydrate beer – production of fermentable glucose
protease	Neutrase®	bacterium *Bacillus*	45 °C pH 6.0 *subtilis*	• brewing – fortifies malt proteases (barley) • baking – softens wheat gluten (e.g. for biscuits)
lactase (β-galactosidase)	Lactozyme®	yeast fungus *Kluyveromyces fragilis*	48 °C pH 6.5	• produces low-lactose milk – suitable for lactose-intolerant people • sweeter so less sugar need be added to drinks • yoghurt – sweeter, longer shelf-life • ice-cream – no lactose crystals, improves scoopability and creaminess • whey – conversion to syrup for use as sweetener
lipase	Lipolase™ (uses DNA technology)	fungus *Aspergillus oryzae*	35 °C pH 7–11 (wide range)	• in detergents – removing fat stains (e.g. frying fat, salad oils, sauces, cosmetics)

QUESTION

Lactase activity is inhibited by the end-product (galactose) it produces. In the industrial processing of whey, the enzyme may be immobilised on cellulose acetate fibres.

- Devise an experiment you can do in the laboratory to illustrate immobilisation.
- How would immobilisation help overcome the problem of end-product inhibition?

An important use of extracted bacterial proteases is in biological detergents. The protease contributes to the breaking down of protein stains when washing clothing. Proteases in biological washing powders include Savinase® and Alcalase®, which are produced by microorganisms, extracted and purified. These enzymes are 'microencapsulated' (surrounded by a soluble inert material) in biological washing powders and become active when dissolved in water. Proteases help to remove stains, such as those from grass, blood and various foods, including egg and gravy. The enzymes hydrolyse the proteins in the stains and the products formed dissolve and disperse in the washing water.

Enzyme immobilisation

Enzymes used in commercial processes are frequently in an **immobilised** form. This means that the enzyme molecules are attached to, or incorporated within, an insoluble material but the enzymes remain active. One of the main advantages of enzyme immobilisation is that the enzyme is easy to separate from the product and so can be re-used; this reduces the overall costs.

DEFINITION

An **immobilised** enzyme is one in which the enzyme molecule has been attached to or enclosed within a porous insoluble material.

There is a wide range of insoluble materials that are used to immobilise enzymes, including alginate gel, porous glass beads, porous alumina, and polyacrylamide entrapment.

Advantages of enzyme immobilisation include the following:

- enzymes can be re-used, saving costs of replacement
- enzymes can be added or removed easily to control the reaction
- the immobilised enzyme can be used in continuous processes
- enzymes are readily separated from the products of reaction
- some enzymes are more stable against heat effects, when they are immobilised.

One important commercial use of an immobilised enzyme is in the production of lactose-reduced milk. This involves the use of immobilised lactase (or β-galactosidase). Sterilised milk flows through a column containing the immobilised enzyme, which hydrolyses the milk sugar, lactose, to its constituent monosaccharides, glucose and galactose. The resulting milk is acceptable to people with lactose intolerance (that is, the inability to digest lactose). The products of lactose hydrolysis taste sweeter than lactose and the resulting sugars are also used in the manufacture of confectionary. (See *Practical: Use of immobilised enzymes* and *Extension material*, p. 43.)

PRACTICAL The effect of temperature on the activity of trypsin

Introduction

Casein is a protein found in milk. When a suspension of casein is hydrolysed, the suspension starts cloudy but becomes clearer as the products dissolve. This hydrolysis is catalysed by proteolytic enzymes such as trypsin. The aim of this experiment is to investigate the effect of temperature on the activity of trypsin, using a suspension of casein as the substrate. Changes in the clarity of the casein suspension will be easier to see if the tubes are checked periodically by holding them against a piece of black card.

Materials

- Casein suspension, 4 per cent (use Marvel® milk powder, 4 per cent solution)
- Trypsin solution, 0.5 per cent
- Distilled water
- Test tubes and rack
- Graduated pipettes or syringes
- Glass beakers or water baths
- Thermometer
- Black card
- Stopwatch

Method

1. Set up a water bath at 30 °C.
2. Pipette 5 cm^3 of casein suspension into one test tube and 5 cm^3 of trypsin solution into another tube.
3. Stand both tubes in the water bath and leave them for several minutes to reach the temperature of the water bath.
4. Meanwhile, set up a control tube containing 5 cm^3 of casein suspension plus 5 cm^3 of distilled water. Stand this tube in the water bath.
5. Mix the enzyme and substrate together and replace the tube in the water bath. Start a stopwatch immediately.
6. Observe the contents of the tube carefully, checking against a piece of black card, and record the time taken for the suspension to become clear.
7. Repeat this procedure at a range of temperatures, for example between 25 °C and 65 °C. Use the same volumes of casein suspension and enzyme solution each time.

Results and discussion

1. Explain the function of the control tube.
2. Describe the relationship between temperature and the time taken for the casein suspension to clear.
3. Find the relative rate of reaction by working out the reciprocal of the time taken for the suspension to clear at each temperature.
4. Plot a graph of the rate of reaction against temperature.
5. Explain your results as fully as you can.
6. What are the sources of error in this experiment? How could it be improved?

PRACTICAL The effect of pH on the activity of catalase

Introduction

Catalase, an enzyme found in many different tissues, catalyses the breakdown of hydrogen peroxide into water and oxygen:

$$2H_2O_2 \rightarrow 2H_2O + O_2$$

Hydrogen peroxide is a toxic substance that can be formed during aerobic respiration and catalase removes this product. The activity of catalase can be measured by finding the rate of oxygen release from hydrogen peroxide.

Minced potato provides a suitable source of catalase and the pH is varied in this experiment using citric acid–sodium phosphate buffer solutions at pH values of 4.4, 5.2, 6.5 and 7.5.

Materials

- Citric acid–sodium phosphate buffer solutions. These are made up as follows. Solution A is 21 g citric acid·H_2O dm^{-3}. Solution B is 28.4 g anhydrous disodium hydrogen phosphate (Na_2HPO_4) dm^{-3}. Buffer solutions of each pH are prepared by mixing solutions A and B in the volumes shown in Table 3.3.

Table 3.3 *Composition of citric acid–sodium phosphate buffer solutions at various pHs*

pH	Solution A / cm^3	Solution B / cm^3
4.4	27.9	22.1
5.2	23.2	26.8
6.5	14.5	35.5
7.5	3.9	46.1

- Hydrogen peroxide solution, 10 volume
- Minced potato
- 5 cm^3 or 10 cm^3 plastic syringe with the end cut off, to measure minced potato
- Graduated pipette or syringe to measure buffer solutions
- Stopwatch

Method

1. Set up the apparatus as shown in Figure 3.12.
2. Place 3 cm^3 of minced potato in the conical flask and add 10 cm^3 of buffer solution. Swirl gently to mix.
3. Replace the bung in the flask ensuring an airtight seal. Fill the graduated tube with water and invert carefully into the beaker of water. Do not place over the end of the delivery tube.
4. Measure 5 cm^3 of hydrogen peroxide into the syringe and connect to the flask. Inject the hydrogen peroxide. This will displace air which must not enter the graduated tube.
5. Immediately position the graduated tube over the end of the delivery tube and measure the volume of oxygen collected every 30 seconds for 5 minutes.
6. Repeat the experiment using a different buffer solution and fresh potato.

Results and discussion

1. Tabulate your results and plot a graph to show the volumes of oxygen evolved against time for each buffer solution.
2. At which pH value was catalase
   (a) most active?
   (b) least active?
3. Which factors were kept constant in this experiment?
4. What were the possible sources of error and how could they be minimised?
5. Explain why changes in pH affect enzyme activity.

Further work

1. This apparatus could be used to investigate other factors affecting catalase activity, such as enzyme or substrate concentration. Enzyme concentration can be varied by using different volumes of potato, for example, 1, 2 and 4 cm^3 of minced potato could be used at pH 6.3, and with a constant concentration and volume of hydrogen peroxide.
2. The substrate concentration can be varied by using 2.5, 5, 10 and 20 volume strengths of hydrogen peroxide. Each time, 5 cm^3 of hydrogen peroxide should be used, at pH 6.3 and with a constant 3 cm^3 of minced potato.

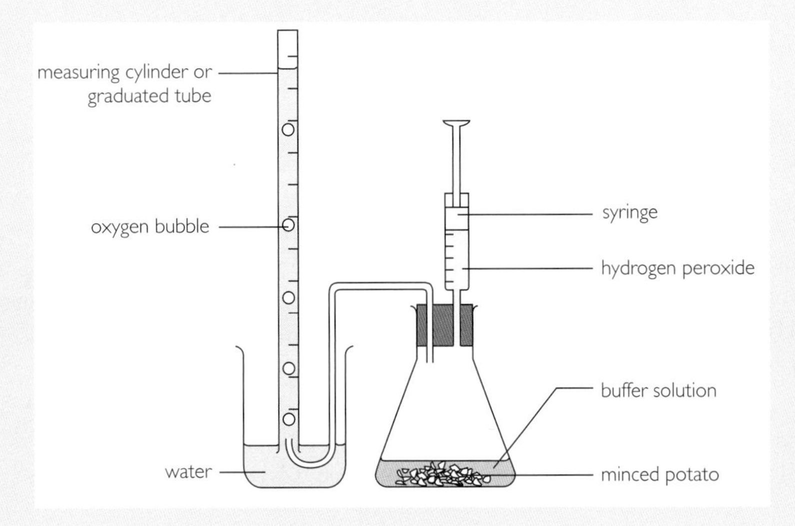

Figure 3.12 Apparatus for the collection of oxygen produced by the decomposition of hydrogen peroxide, catalysed by the enzyme catalase.

PRACTICAL The use of pectinase in the production of fruit juice

Introduction

The purpose of this practical is to investigate the effect of pectinase on the production of apple juice. This practical offers a number of possibilities for individual investigations – for example by studying the effects of enzyme concentration, temperature, types of apple, etc., on the production of juice. You could also investigate the effect of combinations of enzymes, such as cellulase and pectinase, in various proportions.

Materials

- Apples
- Pectinase solution (Pectinex™, available from the National Centre for Biotechnology Education)
- Filter papers (coffee filters work well) and filter funnels
- A knife, or kitchen mincer
- A glass rod
- Two beakers
- Wash bottle of distilled, or deionised, water
- 5 cm^3 plastic syringe
- Two small (e.g. 100 cm^3) measuring cylinders
- Water bath set at 40 °C
- Stopclock

Method

1 Chop the apple into small pieces or mince the apple. Alternatively, you could use prepared apple puree.
2 Place 100 g of apple into each of two labelled beakers (one with enzyme, one without enzyme).
3 Mix 1 cm^3 of Pectinex with 1 cm^3 of distilled water, then add to the appropriate beaker.
4 Add 2 cm^3 of distilled water to the other beaker.
5 Stir the contents of both beakers with a glass rod.
6 Stand both beakers in a water bath at 40 °C and leave for 15 to 20 minutes.
7 Filter the juice from each beaker into measuring cylinders.
8 Record the volumes of juice collected from both lots of apple at intervals of 30 seconds until no more juice is collected.

Results and discussion

CARE!
Do NOT drink the juice.
Handle knives carefully.
Wipe up any enzyme spills and wash the area with water.

1 Record your results in a suitable table, showing the times and volumes of juice collected.
2 Plot a graph of your results, showing the volumes of juice against time.
3 Describe the effect of pectinase on the production of apple juice and suggest an explanation for your results.

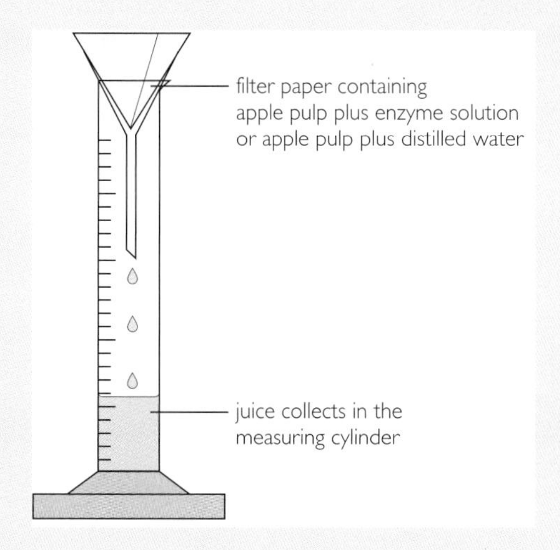

Figure 3.13 Apparatus for investigating the effects of pectinase on the production of apple juice.

PRACTICAL The effect of enzyme concentration on the activity of amylase

Introduction

Amylase is an enzyme that is present in both saliva and pancreatic juice. Its function is to catalyse the hydrolysis of amylose and amylopectin to a mixture of products, including maltose and dextrin. Maltose consists of two α-glucose residues joined by a 1,4 linkage; dextrin is made up of several α-glucose units joined by both 1,4 and 1,6 linkages.

This experiment is to investigate the effect of amylase concentration on its activity. The relative activity of amylase is found by noting the time taken for the starch substrate to be broken down, that is, when it no longer gives a blue-black colour when tested with iodine solution. This time is referred to as the **achromatic point**.

ENZYMES

Materials

- Amylase solution, 0.1 per cent. This must be made up freshly.
- Starch solution, 1.0 per cent. A mixture of equal volumes of the amylase and starch solutions incubated at 35 °C should give a negative result when tested for starch after 2 to 3 minutes. If necessary, the concentration of starch should be increased or decreased.
- Distilled water
- Iodine in potassium iodide solution
- White tile and glass rod
- Graduated pipettes or syringes
- Pasteur pipettes
- Test tubes and rack
- Glass beaker or water bath
- Thermometer
- Stopwatch

Method

1 Prepare four different concentrations of the enzyme solution: undiluted, diluted to a half, diluted to a quarter, and diluted to one tenth of the original concentrations. The volumes of enzyme solution and distilled water to use are shown in Table 3.4.

2 Set up a water bath at 35 °C.

3 Pipette 5 cm^3 of the undiluted enzyme solution into one test tube and 5 cm^3 of starch solution into another test tube. Stand both tubes in the water bath and leave for several minutes to reach the temperature of the water bath.

4 Mix the enzyme and starch solutions together, replace the mixture in the water bath and immediately start a stop watch.

5 At intervals of 1 minute, remove a drop of the mixture and test it with iodine solution on a white tile.

6 Continue the experiment until the mixture fails to give a blue-black colour with iodine solution. Record this as the achromatic point.

7 Repeat this procedure with the other concentrations of amylase. Use exactly 5 cm^3 of enzyme solution and 5 cm^3 of starch solution each time.

Results and discussion

1 Tabulate your results carefully. The relative rate of reaction can be calculated by finding the reciprocal of the time taken to reach the achromatic point at each concentration of amylase.

2 Plot a graph to show the relative rate of reaction against enzyme concentration.

3 Explain your results as fully as you can.

4 What are the sources of error in this experiment? How could it be improved?

Table 3.4 *Composition of amylase solutions made from a 0.1 per cent stock solution*

Volume of amylase solution / cm^3	Volume of distilled water / cm^3	Final amylase concentration
5.0	5.0	0.05
2.5	7.5	0.025
1.0	9.0	0.01

PRACTICAL Use of immobilised enzymes

Introduction

Immobilised enzymes have a wide range of commercial applications, such as their use in the production of lactose-reduced milk, using immobilised lactase (β-galactosidase). Immobilised enzymes are attached to inert, insoluble materials and have a number of advantages over enzymes in free solution, including the ability to re-use the enzyme, which reduces the overall cost of the process. Immobilised enzymes can also be used in continuous processes, which can be automated, and some enzymes are more stable when immobilised and are therefore less likely to be denatured. Enzymes can be immobilised in a range of materials, including agar gels, cellulose and polyacrylamide.

The purpose of this practical is to produce immobilised lactase and to investigate its effect on lactose present in milk.

Materials

- 2 cm^3 lactase, e.g. Novo Nordisk Lactozym®
- 8 cm^3 of 2 per cent sodium alginate solution, made up in distilled water. When making up the alginate solution, add the alginate slowly to warm distilled water and stir constantly.
- 100 cm^3 of 2 per cent calcium chloride solution in a plastic beaker
- Semi-quantitative glucose test strips, such as Diabur 5000
- Small piece of nylon gauze or muslin
- 10 cm^3 plastic syringe
- 10 cm^3 plastic syringe barrel
- Retort stand
- Short length of tubing, to fit plastic syringe, and screw clip
- 100 cm^3 beaker
- Glass rod
- Plastic tea strainer
- Distilled water
- Pasteurised milk

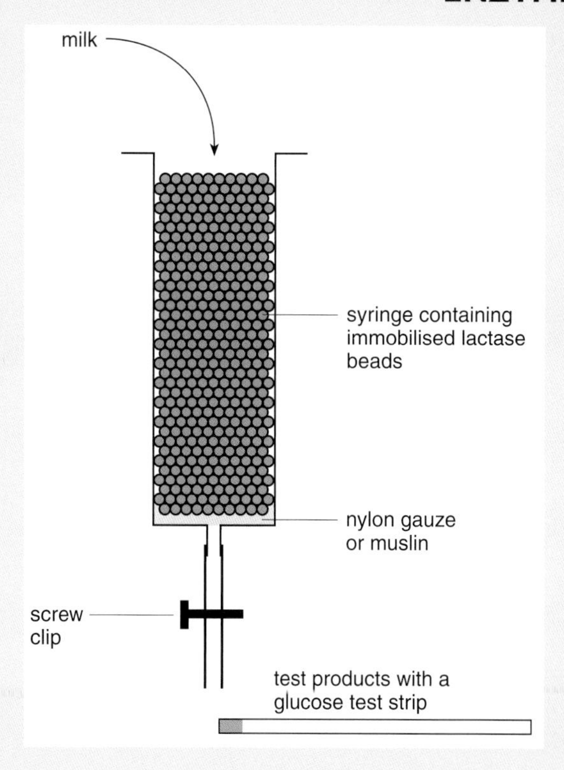

Figure 3.14 Use of immobilised enzymes.

Method

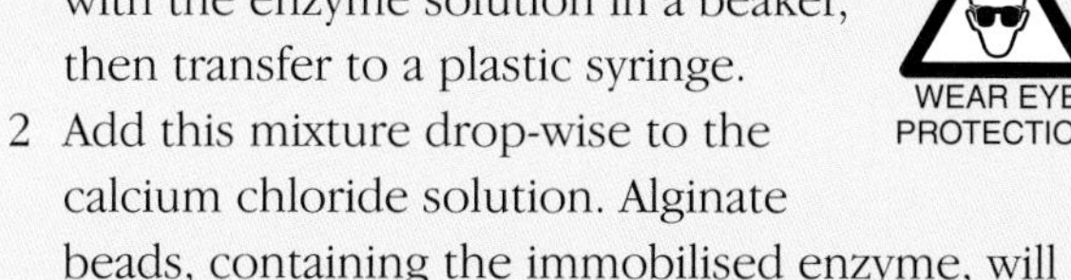

1. Mix the sodium alginate solution with the enzyme solution in a beaker, then transfer to a plastic syringe.
2. Add this mixture drop-wise to the calcium chloride solution. Alginate beads, containing the immobilised enzyme, will form immediately. Leave to harden for 10 to 20 minutes.
3. Strain off the beads using the tea strainer and rinse with distilled water.
4. Put a piece of nylon gauze in a 10 cm^3 syringe barrel, to prevent the beads becoming stuck in the outlet, then add the beads to the syringe. Hold the syringe using a retort stand
5. Close the screw clip, then fill the syringe with milk. Open the clip slightly, and test the products for the presence of glucose. If you use quantitative test strips, such as Diabur 5000, you can investigate the relationship between time at which the sample was taken and glucose concentration.

Further work – immobilised whole cells

Whole cells can be immobilised in a similar way to preparing immobilised enzymes, by entrapping them in alginate beads. Immobilised cells have a number of commercial applications, for example, in the industrial preparation of monoclonal antibodies using hybridoma cells encapsulated in calcium alginate beads.

In this practical, yeast cells are immobilised in calcium alginate and used to hydrolyse sucrose into its constituent monosaccharides.

Materials

- 4 per cent sodium alginate solution in distilled water
- 5 g of fresh yeast mixed with 100 cm^3 of distilled water
- 100 cm^3 of 1.5% calcium chloride solution in a plastic beaker
- 2 per cent sucrose solution
- Glucose test strips, such as Diabur 5000
- Small piece of nylon gauze or muslin
- 10 cm^3 plastic syringe
- 10 cm^3 plastic syringe barrel
- Retort stand
- Short length of tubing, to fit syringe, and screw clip
- 100 cm^3 beaker
- Glass rod
- Plastic tea strainer
- Distilled water

Method

1. Mix 5 cm^3 of the sodium alginate solution with 5 cm^3 of yeast suspension, then transfer the mixture to a plastic syringe.
2. Add this mixture drop-wise to the calcium chloride solution. Leave the alginate beads for 10 to 20 minutes to harden.
3. Strain off the beads and rinse with distilled water.
4. Put a piece of nylon gauze in a 10 cm^3 syringe barrel, then add the beads. Hold the syringe using a retort stand.
5. Close the screw clip, then fill the syringe with 2 per cent sucrose solution. Open the clip slightly and test the products for the presence of glucose.

Suggestions for further work

Immobilised enzymes and whole cells offer a number of possibilities for individual studies. You could investigate the effect of bead diameter, or substrate flow rate on the rate of reactions. Investigate the ability of immobilised yeast cells to utilise a range of carbohydrates in a fermentation experiment. Cells of the one-celled alga ***Chlorella*** can be immobilised and used to investigate the effects of mineral deficiencies on pigment synthesis.

Cellular organisation

Cells, tissues and organs

Living organisms can be distinguished from non-living things by their ability to carry out the characteristic activities of respiration, nutrition, excretion, movement, sensitivity, growth and reproduction. All living organisms are composed of basic units called **cells**. Those organisms consisting of a single cell in which all the characteristic activities take place are often described as **unicellular,** whereas those composed of many cells are described as **multicellular**.

Aggregations of cells

Multicellular organisms vary in their structural complexity, ranging from simple groups, or colonies, of similar cells performing the same activities to complex individuals containing thousands of specialised cells. In the simplest of the colonies, the component cells show little coordination, but as the complexity of organisms increases, so cells become specialised to carry out different functions, which contributes to the efficiency of the whole organism. Cells performing similar functions are organised into **tissues**, and the tissues contribute to the structure of the body **organs**.

The **leaf** of a plant is an organ, composed of several different tissues. Some of these tissues are simple, composed of exactly similar cells performing the same function. Other tissues in the leaf are more complex, with different types of cells carrying out separate functions.

The following tissues comprise a typical mesophytic leaf (Figure 4.1):

- mesophyll – a simple tissue, composed of similar cells adapted for the process of photosynthesis
- palisade mesophyll – composed of palisade cells (long, cylindrical cells, packed with chloroplasts and located just below the upper epidermis)
- spongy mesophyll – composed of cells that are less elongated and that have larger air spaces between them, allowing the circulation and diffusion of gases

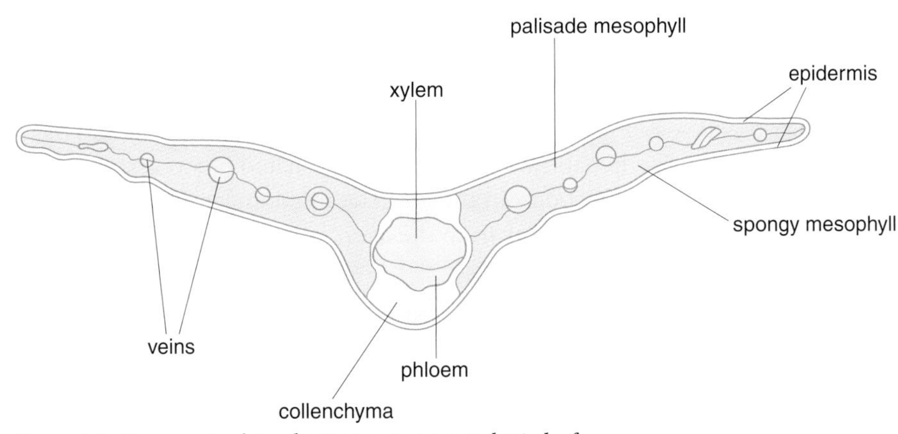

Figure 4.1 Diagram to show the tissues in a mesophytic leaf.

- epidermal tissue – found in the upper and lower epidermis, composed of cells which fit closely together. Amongst these cells are specialised guard cells, that surround the stomatal pores, allowing the exchange of gases between the external environment and the leaf tissues (see *Exchange and Transport, Energy and Ecosystems*, Chapter 1)
- vascular tissue – complex tissue that is comprised of the veins and midrib of the leaf. Vascular tissue includes **phloem** (specialised for the transport of the products of photosynthesis) and **xylem** (through which water and mineral ions are transported).

In addition, a mesophytic leaf may contain other tissues which give support to the leaf:
- collenchyma composed of living cells with extra cellulose thickening: these cells provide flexible support in the midrib region and in the petiole (leaf stalk)
- sclerenchyma made up of lignified fibres, providing rigidity.

Phloem and xylem: Both the phloem and the xylem are made up of several different types of cells, performing different functions.

Phloem tissue contains:
- sieve tubes
- companion cells
- phloem parenchyma
- possibly phloem fibres for support.

Xylem tissue contains:
- vessels and tracheids concerned with the transport of mineral ions and water
- fibres for support
- possibly xylem parenchyma.

In the xylem tissue, the vessels, tracheids and fibres have lignified walls and no cell contents (see page 60, Figure 4.16) (see also *Exchange and Transport, Energy and Ecosystems*, Chapter 2).

The **liver** is an organ composed of liver tissue, together with blood vessels, nerves, lymphatic vessels and fibrous tissue.

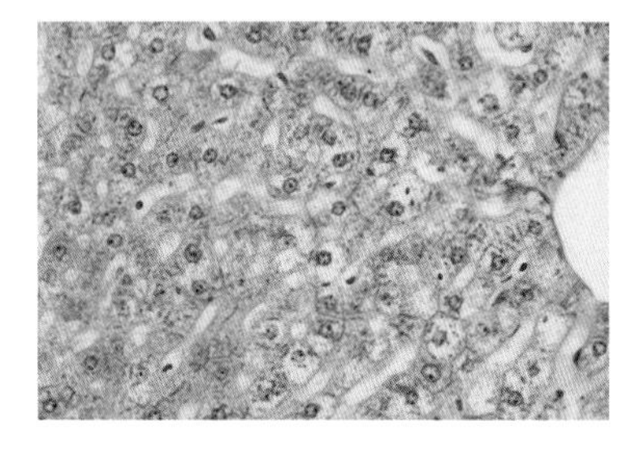

Figure 4.2 Part of a liver lobule, showing rows of hepatocytes, as seen using a light microscope.

Liver tissue is made up of cells, called **hepatocytes**, arranged in vertical cords which radiate in rows around a branch of the hepatic vein called the intralobular vein (Figure 4.2). These cells are all identical, showing no structural or functional differentiation. The cells are packed tightly together and arranged in polygonal blocks called lobules. Associated with these lobules are interlobular blood vessels consisting of branches of the hepatic artery and the hepatic portal vein. The hepatic artery supplies oxygenated blood and the hepatic portal vein brings blood from the gut. The intralobular vessels transport deoxygenated blood away from the liver.

Between the rows of liver cells are blood sinusoids through which blood flows and bathes the liver cells.

One of the functions of the liver cells is the production of bile, which is secreted into bile canaliculi. The bile passes from the canaliculi into bile ducts, which transport the bile to the gall bladder, where it is stored.

Plant and animal cells

The liver cell and the palisade cell are examples of cells specialised for different functions, but they have several features in common, as well as features that distinguish plant cells from animal cells (Figure 4.3).

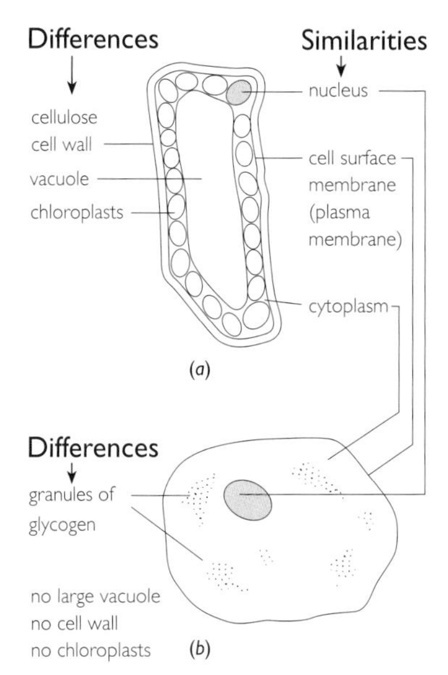

Figure 4.3 Comparison of (a) a plant and (b) an animal cell as viewed with a light microscope.

Features common to both types of cell are:

- a nucleus, which contains the genetic material in the form of **chromosomes**
- the cytoplasm, containing a solution of ions and organic compounds
- the cell surface membrane, which forms a selective barrier between the cell and its external environment.

Plant cells have non-living **cellulose cell walls** and often contain large fluid-filled vacuoles in the cytoplasm. In addition, plant cells that photosynthesise possess green, disc-shaped structures in the cytoplasm, called **chloroplasts**. Animal cells do not possess cell walls or chloroplasts, and rarely have large vacuoles.

Eukaryotic cells

Most cells are fairly small structures, ranging in size from 10 to 150 μm (0.01 to 0.15 mm) in diameter, and so far we have only considered those features of cells that are visible using the light microscope. Under the low power of the light microscope, a magnification of about 100 times is achieved, rising to about 400 times or just above at high power, depending on the lenses used. Electron microscopy (Figure 4.4) enables much greater magnification, from about 1000 to 200 000 times, and reveals the fine structure, or **ultrastructure**, of the cell.

> **DEFINITION**
>
> **Eukaryotic cells** have linear DNA contained in a nuclear membrane, organelles that are membrane-bound, and dense ribosomes.

The nucleus is seen to be surrounded by a double membrane, the **nuclear envelope**, and the cytoplasm appears as a complex system of membranous sacs, the **endoplasmic reticulum**. It is possible to discover the presence and nature of other small structures, such as **mitochondria**, **ribosomes** and **Golgi apparatus**, within the cytoplasm. It is usual to refer to these structures contained within the cell as organelles. Some of them, such as mitochondria, are surrounded by membranes, while others, such as ribosomes, are not (Figures 4.5 and 4.6).

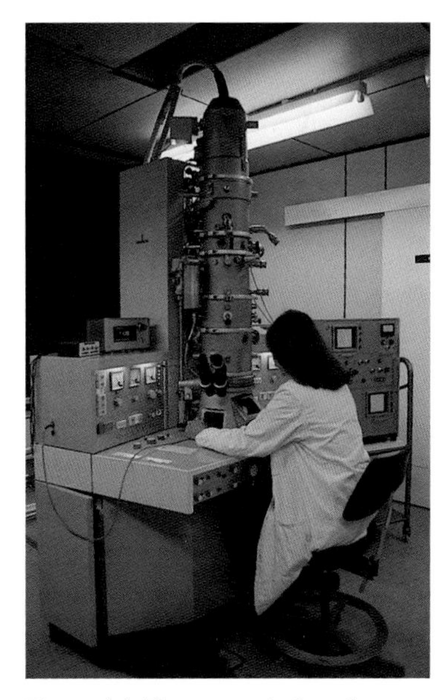

Figure 4.4 The transmission electron microscope (TEM) is able to resolve structures less than 2 nm across, 100 times better than the best light microscopes.

Cell surface membrane

The living material of all cells is surrounded by a cell surface membrane, sometimes referred to as the **plasma membrane**. This membrane forms a selective barrier between the cell contents and the external environment. It controls the passage of substances into and out of the cell, regulating the internal environment and providing suitable conditions for the chemical reactions that take place inside the cell. The membranes of the endoplasmic reticulum and the Golgi apparatus, together with the membranes surrounding the nucleus, appear to be similar in structure to the cell surface membrane.

As can be seen in Figure 4.8, the cell surface membrane consists mainly of lipids and proteins. Carbohydrates are also present, but are always found in association with lipids and proteins, as glycolipids and glycoproteins respectively. The nature and general properties of these groups of molecules are discussed in Chapter 1, but it is relevant here to mention some special features that are associated with the structure and properties of membranes.

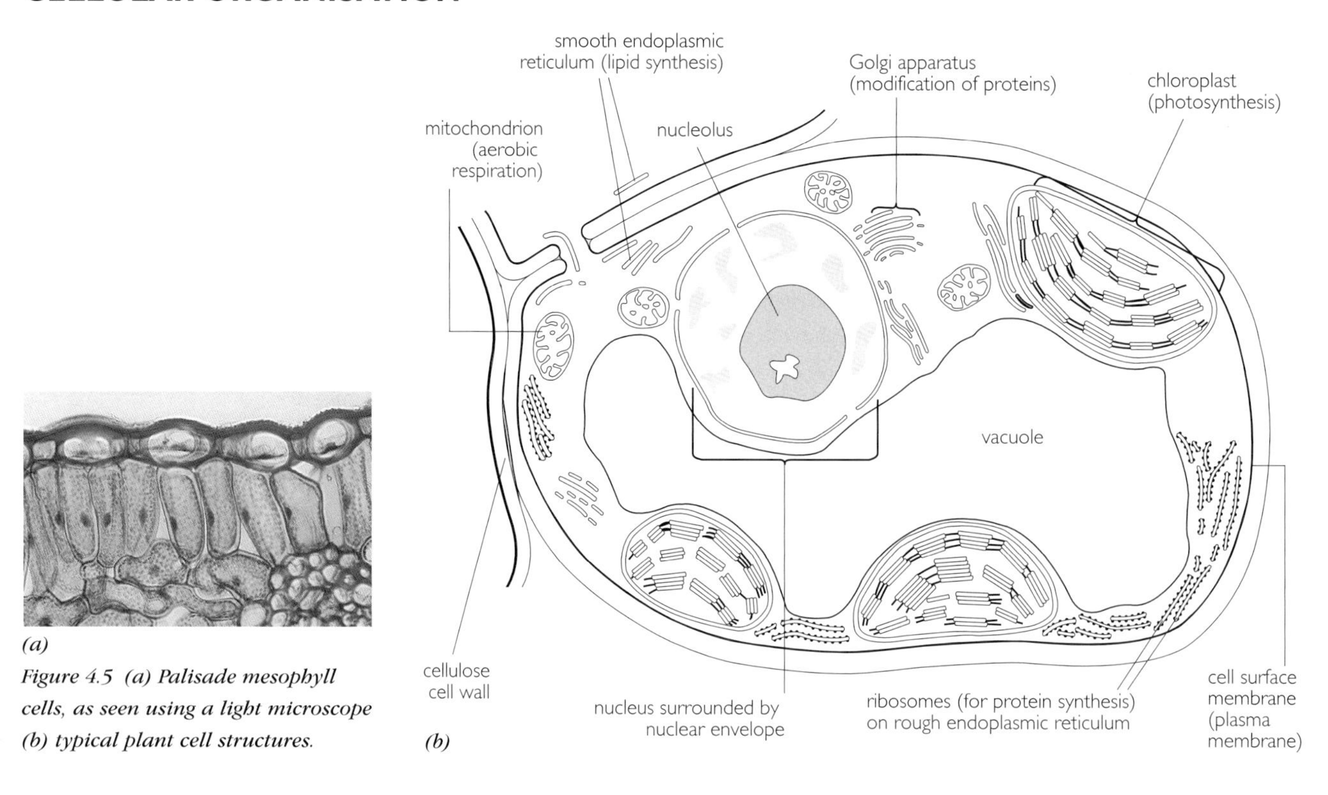

Figure 4.5 (a) Palisade mesophyll cells, as seen using a light microscope (b) typical plant cell structures.

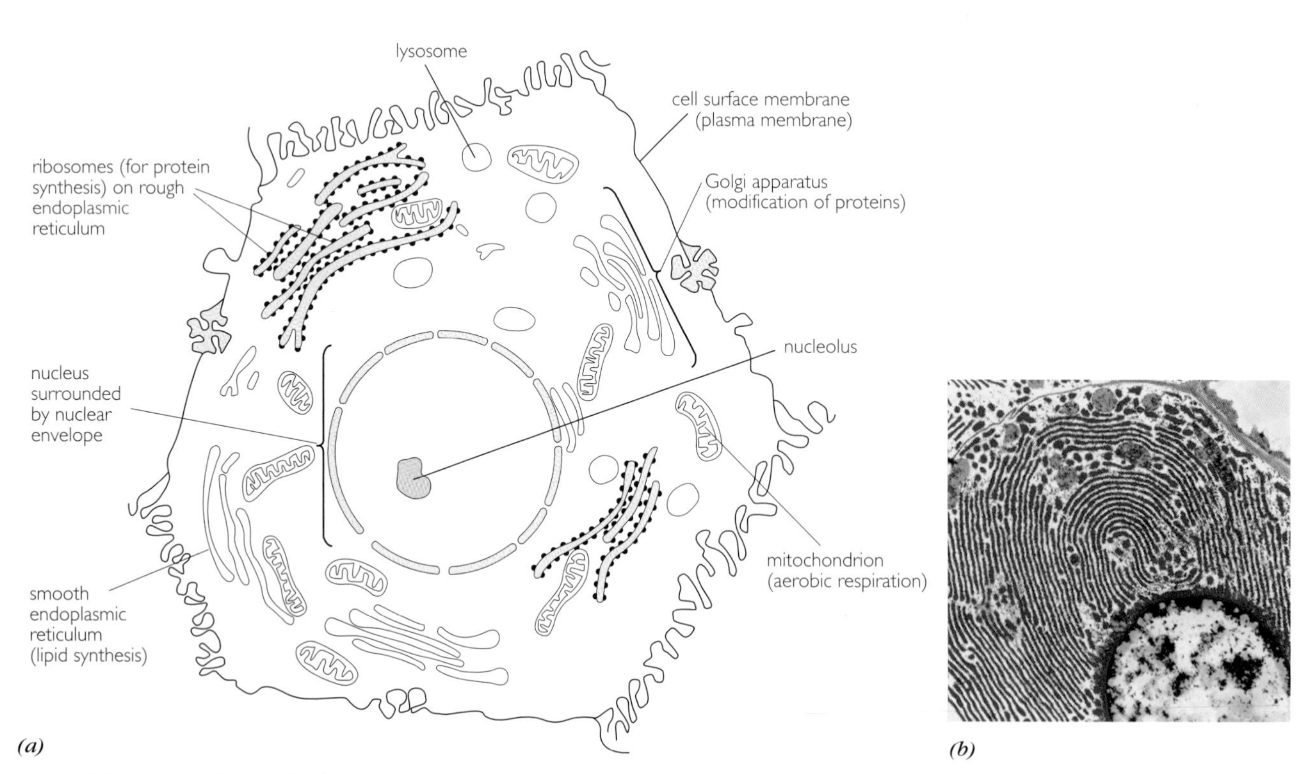

Figure 4.6 (a) Typical animal cell structures; (b) electronmicrograph of a human liver cell.

Background Information

Lipid and protein molecule arrangement

Before the use of electron microscopy, the membrane was known to consist of lipid and protein molecules, but their arrangement was not known. In 1935, Davson and Danielli proposed that the membrane consists of two layers of lipid molecules, a lipid bilayer, coated on both surfaces with a layer of protein molecules (Figure 4.7). They had calculated that the thickness of the membrane would be about 7.5 nm (1 nm = 1/1000 μm) and early electronmicrographs appeared to confirm this structure.

With improvements in electron microscopes and the use of different techniques in the preparation of material to be examined, it was possible to obtain more detailed information, and in 1972 Singer and Nicolson put forward the 'fluid-mosaic' model of membrane structure (Figure 4.8), suggesting that the membrane is a fluid structure around the cell with a mosaic of different proteins in it. This model incorporates the ideas of Davson and Danielli, and others, in that there is a lipid bilayer, but it suggests that, in addition to the protein molecules that are embedded in the bilayer, some larger protein molecules span the membrane. It is also suggested that the lipid and protein molecules are able to move about, accounting for the ease with which membranes fuse with each other.

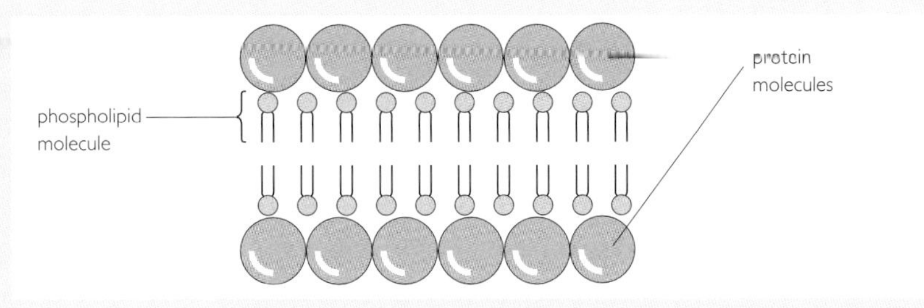

Figure 4.7 Davson and Danielli's 1935 model for cell surface membrane structure proposed that the lipid layer is coated with globular proteins.

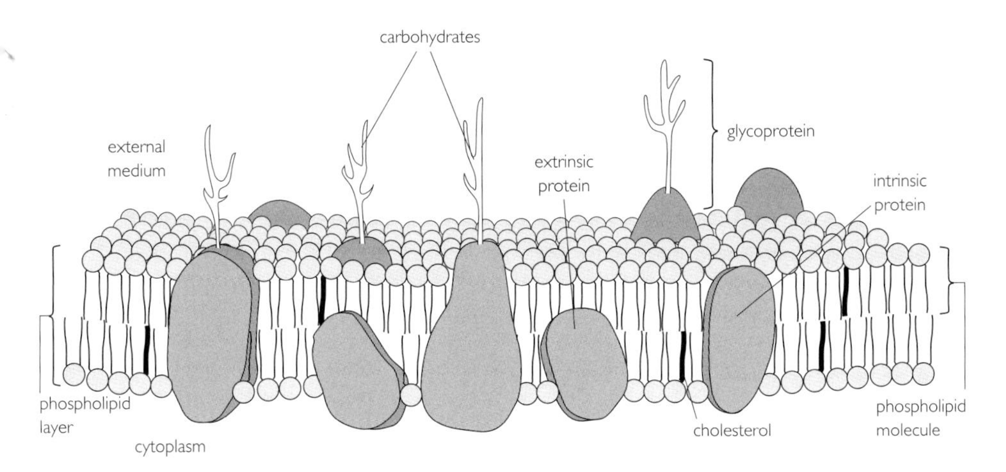

Figure 4.8 Singer and Nicolson's 1972 fluid-mosaic model proposed that proteins float in the lipid bilayer.

Lipids

Three types of lipids are found in membranes:

- **phospholipids**, which contain a phosphate group and form the bilayer
- **glycolipids**, which have a branching carbohydrate (polysaccharide) molecule and act as recognition sites for other cells
- **cholesterol**, which contributes to the structure.

In both phospholipids and glycolipids, one part of the molecule associates with water (**hydrophilic**) and the other part does not mix with water (**hydrophobic**). The hydrophilic portion is said to be **polar** and the hydrophobic portion **non-polar**. In phospholipids, which are the most common lipids in the membrane, the part of the molecule containing the phosphate group is referred to as the polar head and the two fatty acid chains form the non-polar tails. In a bilayer of phospholipid molecules, the non-polar tails face inwards and the polar heads face outwards.

Glycolipids are less common in the membrane, but where present they always occur in the outer layer with the carbohydrate portions, forming the **glycocalyx**, and extending outwards into the intercellular space. Cholesterol molecules have a different structure, but do have polar and non-polar regions. They are arranged in the bilayer with their polar groups close to the polar groups of the other lipid molecules.

Proteins

A large number of different proteins can occur in cell membranes. As shown in Figure 4.8, some completely span the membrane (**intrinsic**), while others occur embedded in one half or located on the inner surface (**extrinsic**). Like the membrane lipids, the proteins have polar and non-polar regions. Interactions between the hydrophobic and hydrophilic regions of the proteins and the lipids help to maintain the stability of the membrane. Glycoproteins are common in membranes, the branching carbohydrate portions contributing to the glycocalyx mentioned above. As with the glycolipids, they are always found in the outer layer.

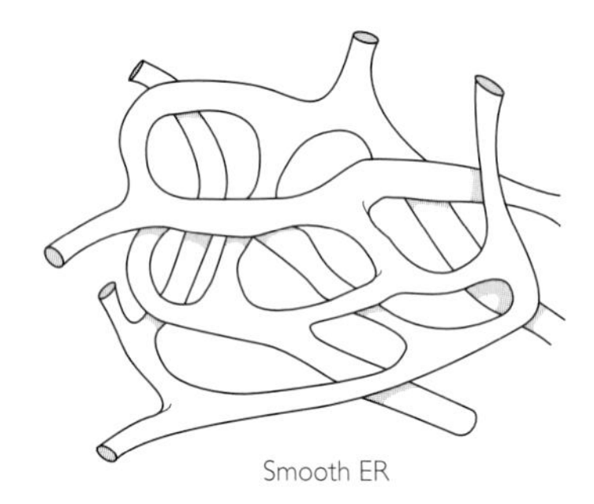

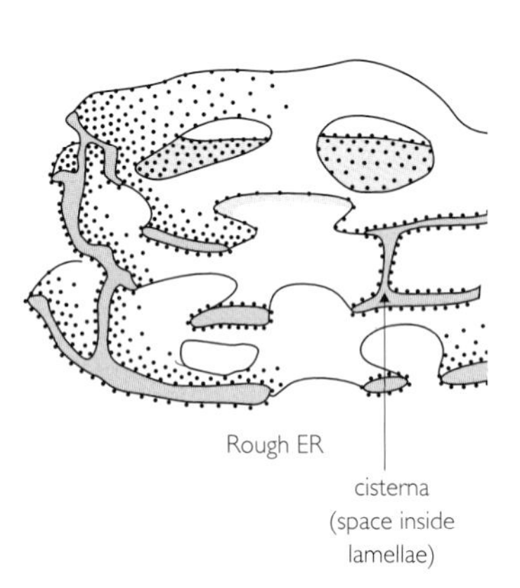

Figure 4.9 Endoplasmic reticulum (ER), showing smooth and rough types.

Endoplasmic reticulum

The endoplasmic reticulum (Figure 4.9) is made up of a complex system of membrane-bound flattened sacs or tubules, called **cisternae**. Where ribosomes are present on the outer surface of the membranes, it is referred to as **rough endoplasmic reticulum** (**RER**), and where there are no ribosomes, it is called **smooth endoplasmic reticulum (SER)**. The cisternae of the SER are usually more tubular than those of the RER. The **ribosomes** are involved in the synthesis of proteins, which are transported in the RER. The RER appears to be extensive in cells that actively make and secrete proteins. The SER is concerned with the synthesis of lipids and is well developed in cells that produce steroid hormones and in liver cells.

Golgi apparatus

The Golgi apparatus, or Golgi body, consists of a stack of flattened cisternae and associated vesicles. It is present in all cells, but appears to be more prominent in those that are actively producing enzymes and other secretions. It is involved in the modification of proteins synthesised by the ribosomes (Figure 4.10).

Small membrane-bound cavities, or vesicles, are pinched off the endoplasmic reticulum and fuse with the cisternae of the Golgi apparatus. The vesicles from the RER contain proteins, which then have carbohydrate molecules attached to them, resulting in the formation of glycoproteins, such as are found in the cell surface membrane. Vesicles containing these modified proteins bud off from

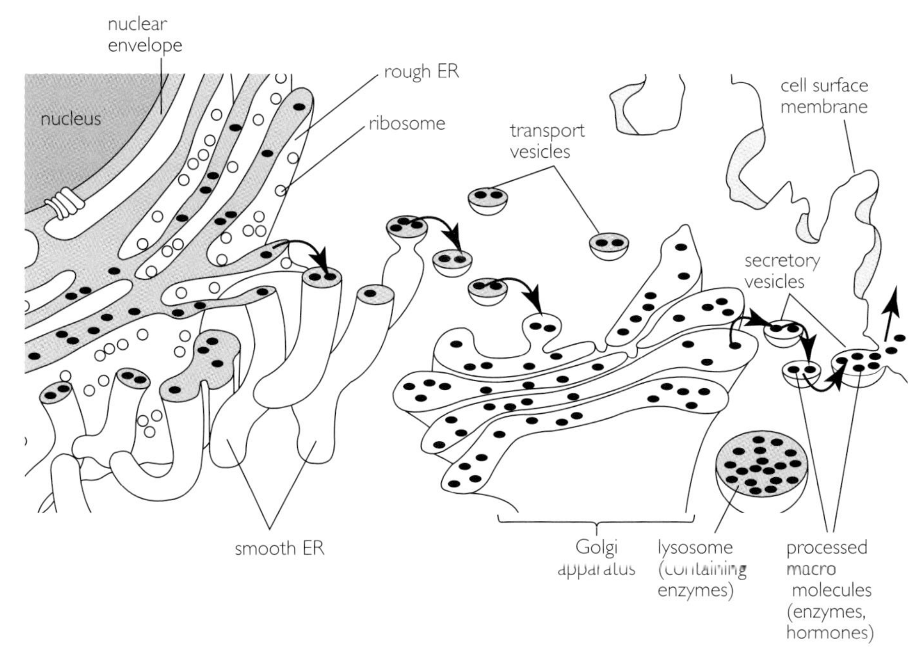

Figure 4.10 Golgi apparatus, showing its relationship with the endoplasmic reticulum (ER) and its role in packaging materials for export from a secretory cell.

the cisternae and can be secreted from the cell. The vesicles fuse with the cell surface membrane, releasing their contents. It has been possible to trace this pathway by using radioactively labelled amino acids and following the formation of proteins, their modification to glycoproteins and subsequent release from the cell.

The Golgi apparatus has also been shown to be involved with the transport of lipids within cells and plays an important role in the formation of **lysosomes**, which are membrane-bound organelles containing digestive enzymes. Lysosomes have a number of different functions within cells, including the breakdown of unwanted structures, such as old mitochondria and other worn-out organelles.

Nucleus and nuclear envelope

The nucleus of a cell is easily seen using a light microscope and appears as a dense, spherical structure within the cytoplasm, from 10 to 20 μm in diameter, surrounded by a membrane. Electron microscopy reveals this membrane as a double membrane, forming the **nuclear envelope**, which separates the nuclear contents from the rest of the living material of the cell (Figure 4.11). The outer membrane is continuous with the membranes of the endoplasmic reticulum and bears ribosomes. The space between the two membranes is very small (from 20 to 40 nm wide) and is continuous with the cisternae of the endoplasmic reticulum. The inner and outer membranes fuse at intervals giving rise to the **nuclear pores**.

The nuclear envelope enables the regulation of the movement of molecules between the nucleus and the cytoplasm and helps to separate the reactions taking place in the nucleus from those taking place in the cytoplasm. Within the nucleus is the **nucleoplasm**, a jelly-like material containing the

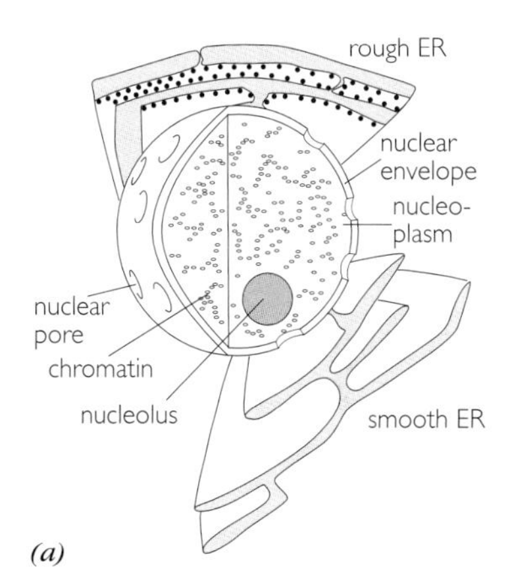

(a)

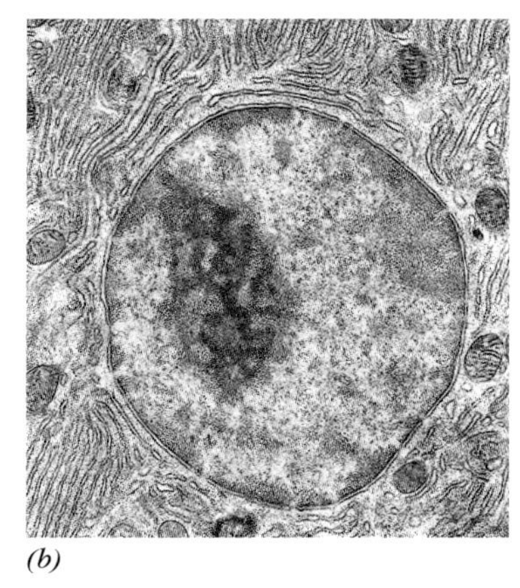

(b)

Figure 4.11 (a) Cell nucleus, showing internal structure and relationship with the endoplasmic reticulum (ER); (b) electronmicrograph of cell nucleus.

chromosomes and one or more **nucleoli**. The chromosomes contain the hereditary material in the form of **DNA (deoxyribonucleic acid)** attached to proteins called **histones**, and are only visible when the nucleus is undergoing division. The nucleoli, which are not surrounded by membranes, can be seen as circular, granular structures. Their function is to make rRNA **(ribonucleic acid)** and to assemble ribosomes. During the early stages of nuclear division, the nucleoli in a cell disperse and are no longer visible, reappearing during the later stages.

In a nucleus that is not undergoing division, the chromosomes form a diffuse network referred to as **chromatin**, so called because it stains easily with dyes and can be readily observed under the microscope.

The function of the nucleus is to control activities within the cell by controlling the chemical reactions. This is achieved by regulating the synthesis of proteins and enzymes. When a cell undergoes division, the nucleus divides first, thus ensuring that the new cell has an exact copy of the information contained in the chromosomes.

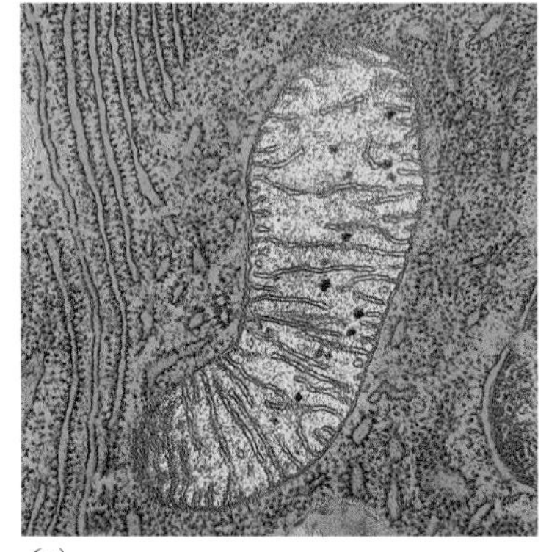

(a)

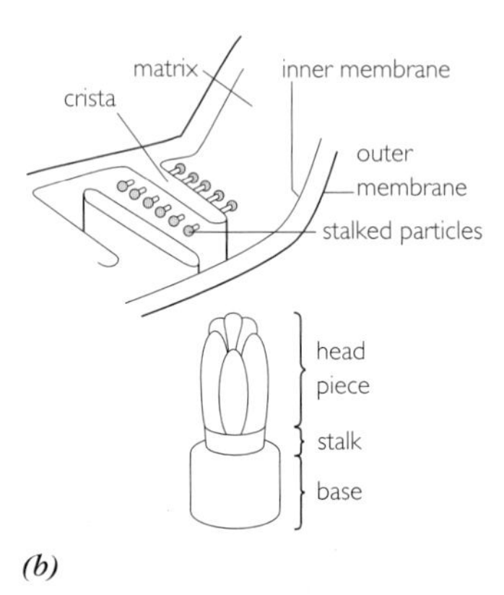

(b)

***Figure 4.12** (a) Mitochondrion, showing internal structure; (b) structure of individual stalked, or elementary, particle.*

Mitochondria

Mitochondria are barely visible under the light microscope, but electron microscopy indicates that they are rod-shaped, up to 1 μm wide and typically about 7 μm long (Figure 4.12). Each mitochondrion is surrounded by a double membrane, or envelope. The outer membrane is smooth, but the inner is folded into many shelf-like projections called **cristae** (singular: crista). The inner membrane encloses the mitochondrial **matrix**, which has a jelly-like consistency.

Mitochondria are the sites of the reactions of aerobic respiration within cells. The details of this process are not required for AS. However, it is relevant to note here that the matrix contains enzymes involved with the **tricarboxylic acid (TCA) cycle**, alternatively known as the **Krebs cycle**. This cycle is named after Sir Hans Krebs, who first discovered the sequence of reactions. The reactions in which **ATP (adenosine triphosphate)**, the energy currency of cells, is produced take place on the cristae.

Special staining techniques used on isolated fragments of the inner membrane have shown that there are structures, called **elementary particles**, on the matrix side of the cristae. These particles, which contain the enzyme involved in the synthesis of ATP, appear as tiny spheres with a diameter of about 9 nm, on stalks 4 nm high, and are spaced along the membrane at regular intervals. The elementary particles only become visible as stalked spheres when the membrane structure is disturbed, as would happen during the breaking up and preparation of mitochondria for viewing by electron microscopy.

Present in the matrix there are mitochondrial ribosomes, which are smaller than those found in the cytoplasm of the cell, and mitochondrial DNA in the form of a circular molecule.

Chloroplasts

Chloroplasts occur in the cells of the photosynthetic tissue of plants. They belong to a group of organelles known as **plastids**, which often contain pigments. Chloroplasts occur in large numbers in the palisade cells of the leaves of flowering plants. They are disc-shaped structures and appear green due to the presence of the pigment **chlorophyll**. They range from 2 to 5 μm in diameter and are 1 μm thick, easily seen using a light microscope. Electron microscopy shows that each chloroplast is surrounded by a double membrane, the **chloroplast envelope**, enclosing the stroma in which there is a system of flattened membranous sacs called **thylakoids**, or **lamellae**. **Grana** are formed from several thylakoids, or lamellae, stacked together in the matrix (Figure 4.13).

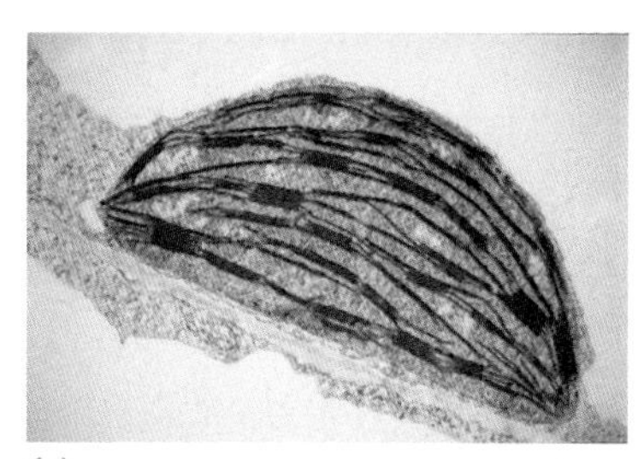

(a)

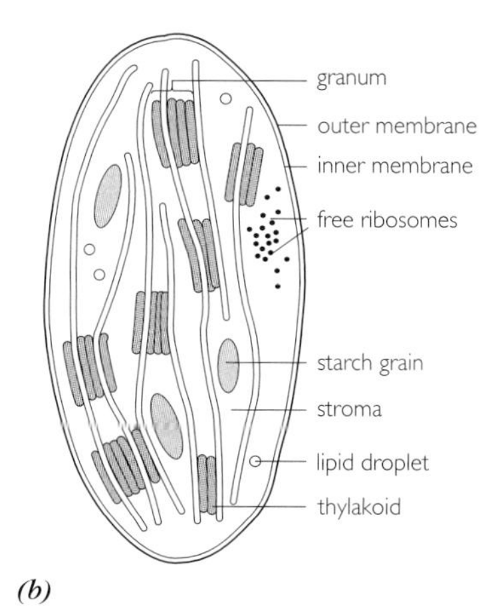

(b)

Figure 4.13 (a) Electronmicrograph of chloroplast; (b) chloroplast from leaf palisade cell, showing internal structure.

The membrane system formed by the thylakoids is the place where the **light-dependent** reactions of photosynthesis occur. The chlorophyll molecules, whose function is to trap the light energy, are situated on the thylakoids. The stroma contains the enzymes necessary for the **light-independent** reactions in which carbon dioxide is converted into carbohydrates.

Also present in the stroma are small ribosomes and circular DNA molecules, starch grains and lipid droplets.

Microtubules

Microtubules are very fine, tubular organelles, which contribute to the complex network of fibrous proteins making up the **cytoskeleton** in the cytoplasm of living cells. They are straight, unbranched, hollow structures, which vary in length, but have an external diameter of about 20 to 25 nm. The walls of these tubules, estimated to be about 5 nm thick, are composed of subunits of the protein **tubulin**. They can increase in length by the addition of more subunits at one end, and shorten by their removal, so they may be constantly built up and broken down within cells.

Microtubules contribute to the structure of other cell organelles, including **centrioles**, and make up the **spindle** in cells undergoing nuclear division. They are involved with the following activities within cells:

- determination and maintenance of shape
- transport of granules and vesicles within the cytoplasm
- movement of chromosomes during nuclear division.

Centrioles

Centrioles are present in most animal cells and in the cells of other organisms, such as **fungi** and some **algae**. They are hollow cylindrical organelles with a diameter of 0.15 μm and length 0.5 μm. The wall of each centriole is made up of nine triplets of microtubules arranged at an angle, as shown in Figure 4.14. Where they are present in cells, they occur in pairs, arranged at right angles to each other, forming the **centrosome**. They are often situated close to the Golgi apparatus and appear to have a role in the organisation of the spindle in animal cells.

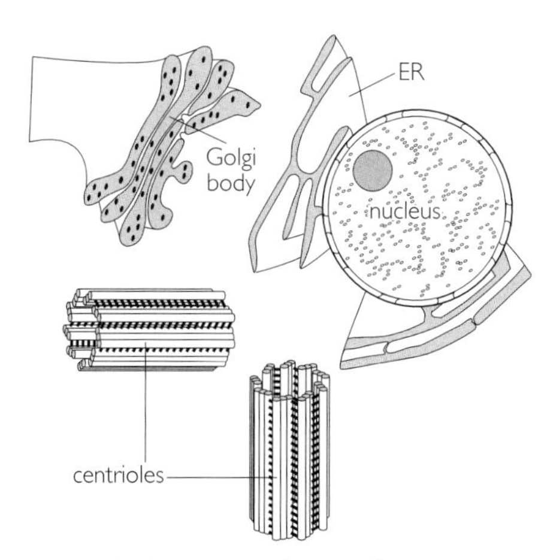

Figure 4.14 Position of centrioles in relation to other organelles.

Cell walls

Cell walls surround the living contents of cells, but are themselves non-living so they are not classed as organelles. They are relatively rigid structures secreted by the living material and provide support and protection for the cell contents. In this unit, we are mainly concerned with the **cellulose cell walls** of green plants, but cell walls are also present in other groups of organisms, such

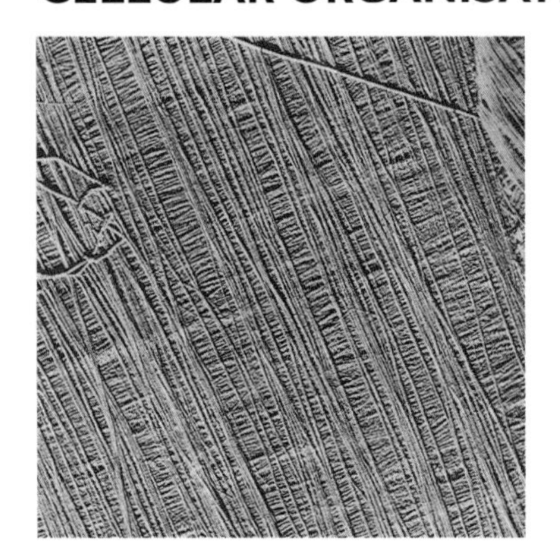

Figure 4.15 Electronmicrograph of cell wall from a plant, showing criss-crossing cellulose microfibrils (× 10 000).

as the fungi, some of the **Protoctista** (algae and protozoa) and the **Prokaryotae (bacteria)**.

In green plants, the first wall to be laid down following cell division is called the **primary wall** and it consists of cellulose **microfibrils** embedded in a **matrix** of complex **polysaccharide** molecules, which include **pectins** and **hemicelluloses.** Each cellulose microfibril is made up of about 2000 cellulose molecules cross-linked to each other to form a bundle. In this primary wall, the microfibrils run in all directions, allowing for the growth and stretching of the wall. In most cells, after the maximum size has been reached, additional cellulose microfibrils are laid down, building up a **secondary wall** (not to be confused with **secondary thickening**, which involves the addition of new cells for example, in the stems of some plants). The microfibrils making up each of these additional layers are usually orientated at the same angle, with each subsequent layer orientated at a slightly different angle to the one below (Figure 4.15). Palisade cells in the leaf do not normally develop secondary walls, but in other cells, such as those forming the tissue **collenchyma**, the additional layers of cellulose microfibrils can be quite thick.

The cell walls of neighbouring cells are held together by the **middle lamella**, a sticky jelly-like substance containing a mixture of **magnesium** and **calcium pectates.**

Cell walls have narrow pores through which very fine strands of cytoplasm called **plasmodesmata** (singular: **plasmodesma**) pass. The plasmodesmata range from 100 to 500 nm in diameter, and provide a connection between the living contents of adjacent cells and a pathway for the movement of material from cell to cell.

The main functions of the cell wall are:

- to provide mechanical strength and support to the cell
- to resist expansion when water enters

The cellulose microfibrils have high tensile strength, which makes the cell walls mechanically strong. The matrix contributes to the strength by improving the resistance to shearing and compression forces, as well as spreading out the microfibrils and protecting them from abrasion and chemical attack. The construction of the plant cell wall has been compared to that of reinforced concrete, where the steel rods are equivalent to the microfibrils and the concrete to the matrix.

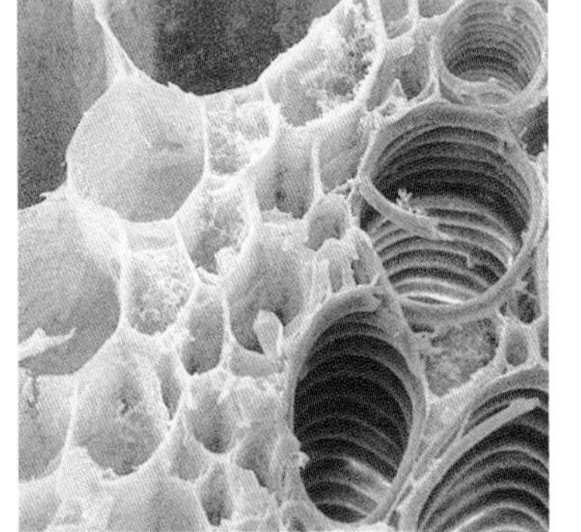

Figure 4.16 Photomicrograph of typical xylem vessels, showing stained lignin spirals, and rings.

The secondary walls of some of the cells in tissues such as **xylem** (Figure 4.16), undergo a process of **lignification**, where a complex molecule called **lignin** is deposited amongst the cellulose layers. The lignin holds the cellulose microfibrils together and makes the cell walls hard and rigid. The tissues in which lignification has occurred have extra tensile strength and a greater resistance to compression, making them ideal as construction materials. Cellulose cell walls are freely permeable to water but, where lignification occurs, the lignified areas are impermeable.

Prokaryotic cells

The terms prokaryotic and eukaryotic are derived from the Greek *karyon* (meaning 'kernel' or 'nucleus'), together with the prefix *pro* (meaning 'before') and *eu* (meaning 'true'). These terms were first used in 1937 by the French marine biologist Edouard Chatton. Both prokaryotic and eukaryotic cells carry out the same activities characteristic of living organisms and so share some common features, but there are significant differences in their internal organisation.

The cells of the members of the Kingdom Prokaryotae, which includes all the bacteria, are simple in structure and lack complex organelles and internal membranes (Figure 4.17). They are referred to as **prokaryotic**. Cells that are typical of plants and animals have a complex system of internal membranes and organelles. In these cells, the nucleus is surrounded by a nuclear envelope which separates the genetic material from the rest of the cytoplasm. These cells are termed **eukaryotic**.

The first representatives of the Prokaryotae are thought to have evolved about 3500 million years ago, and it has been suggested that eukaryotic cells evolved from prokaryotic cells about 1500 million years ago.

Most prokaryotic cells range in length from 1 to 10 μm with a diameter no greater than 1 μm. The cell wall is rigid, the main component being a **peptidoglycan** called **murein** and not cellulose as in plant cell walls. Both cellulose and murein are polysaccharides, but in murein the parallel polysaccharide chains are linked by short peptides to form a complex three-dimensional network. Some bacteria have a capsule or slime layer on the outside of the cell wall.

In prokaryotic cells, the DNA is arranged as a single closed loop, sometimes called a **chromosome** but it is not associated with proteins as true chromosomes are. In addition to the chromosome, there may be smaller loops of DNA, called **plasmids**, present in the cytoplasm. There are ribosomes present throughout the cytoplasm. These are concerned with protein synthesis and are of the 70S type.

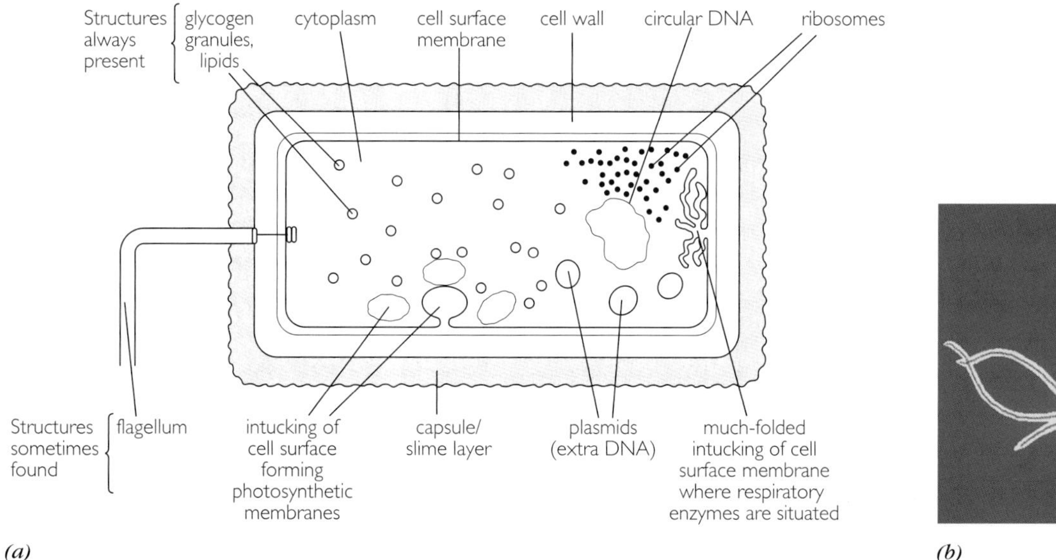

(a)

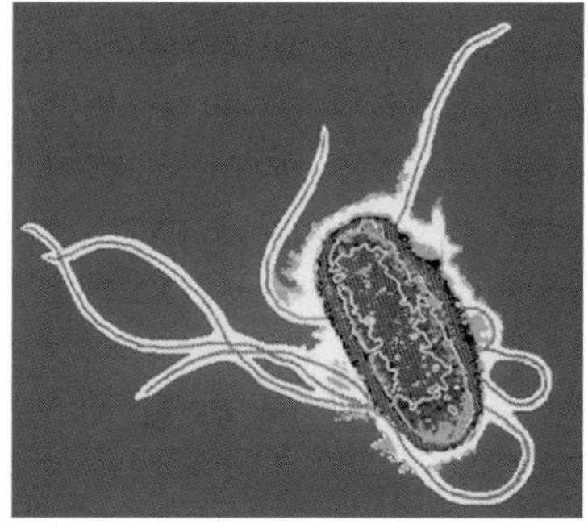

(b)

Figure 4.17 (a) Typical prokaryotic cell, showing structures always or sometimes present (b) electronmicrograph of Escherichia coli.

In prokaryotes, although there is no complex system of intracellular membranes, some of them have a cell surface membrane which forms extensively folded regions in the cytoplasm. A good example of this is seen in the photosynthetic bacteria, where the chlorophylls and enzymes needed are found on **thylakoids** formed by intuckings (invaginations) of the cell surface membrane. Similarly, in some bacteria, the enzymes associated with aerobic respiration are located on a folded structure formed by invagination of the cell surface membrane. Filamentous structures called **pili**, or **fimbriae**, may be present attached to the cell wall or capsule. These can be up to 1 μm long and are usually less than 10 nm thick. Their function is concerned with cell-to-cell attachment or cell-to-surface attachment. **Flagella** are found in some groups of bacteria: they may be all over the cell, or limited to a group at one or both ends of the cell, or present singly. They consist of a single fibril, usually about 20 nm thick and up to several micrometres in length, rather than the 9+2 arrangement of microtubules characteristic of eukaryotes. The roles of these structures are summarised in Table 4.1.

The differences in the structure and organisation of prokaryotic and eukaryotic cells are summarised in Table 4.2.

Table 4.1 *Roles of structures in prokaryotic cells*

Structure	Role
cell wall	confers rigidity and shape; prevents cell from swelling and bursting; permeable to water, ions and small molecules but not to proteins or nucleic acids
cell surface (plasma) membrane and its invaginations	partially permeable membrane forming a selective barrier between cell contents and external environment; controls passage of substances into and out of the cell; regulates internal environment
thylakoids (in photosynthetic bacteria)	invaginations where chlorophylls and enzymes for photosynthesis are located
invaginations linked with respiration	location of enzymes associated with aerobic respiration
flagella	present in motile bacteria; enable movement in response to certain stimuli
bacterial chromosomes	contain genetic information
plasmids	carry genes which help bacteria to survive in adverse conditions; able to replicate independently of main chromosome
glycogen granules and lipid droplets	food reserves

Table 4.2 *Table of differences between prokaryotic and eukaryotic cells*

Prokaryotic cells	Eukaryotic cells
mostly small cells ranging in size from 1 to 10 μm	cells bigger, typically 10 to 150 μm can be up to 400 μm
rigid cell wall containing murein present	cell walls when present contain cellulose (green plants) or chitin (fungi)
no true nucleus	true nucleus present surrounded by nuclear envelope
circular DNA; no true chromosomes	linear DNA with associated proteins forming true chromosomes
no nucleolus	nucleolus present
no endoplasmic reticulum	endoplasmic reticulum present with associated Golgi apparatus, lysosomes and vacuoles
smaller (70S) ribosomes	larger (80S) ribosomes
no membrane-bound organelles; lack mitochondria; lack chloroplasts, photosynthetic membranes (thylakoids) in photosynthetic bacteria	many membrane-bound organelles: chloroplasts with lamellae in photosynthetic organisms; mitochondria for aerobic respiration
flagella, when present, lack microtubules	flagella have 9+2 arrangement of microtubules

EXTENSION MATERIAL

Viruses

Viruses are much smaller than bacteria and cannot be seen using a light microscope (Figure 4.18). They range in size from about 20 nm to 400 nm and do not have a cellular structure, so they are described as **akaryotic**. They are intracellular parasites of plants, animals and bacteria, totally dependent on their host cells. The only characteristic they have in common with other living organisms is that they can reproduce once they are inside their host's cells. Viruses do not respire, feed, excrete, move, grow or respond to stimuli. They disrupt the normal activities of cells, often with harmful effects on the host organism, so they are associated with disease.

A virus consists of:

- a core of nucleic acid
- a protein coat, or capsid.

The nucleic acid may be:

- double-stranded DNA, as in Herpes simplex, which causes cold sores
- single-stranded DNA, as in Parvovirus, which causes gastroenteritis
- single-stranded RNA, as in the influenza virus and the human immuno- deficiency virus (HIV).

Most of the viruses causing diseases in plants, such as the tobacco mosaic virus, contain RNA.
The capsid surrounds the nucleic acid and consists of a number of subunits called **capsomeres**, arranged to form a geometrical structure. In addition, some viruses have an outer envelope of carbohydrate or lipoprotein and many bacteriophages (bacterial viruses) have tails that form part of the mechanism by which they gain entry to host cells (Figure 4.19).

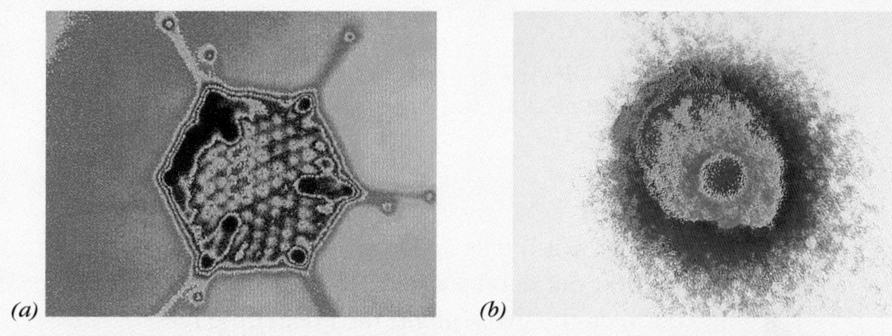

Figure 4.18 Electronmicrographs of (a) bacteriophage T2, which parasitises the gut bacterium Escherichia. coli, *magnified; (b) tobacco mosaic virus.*

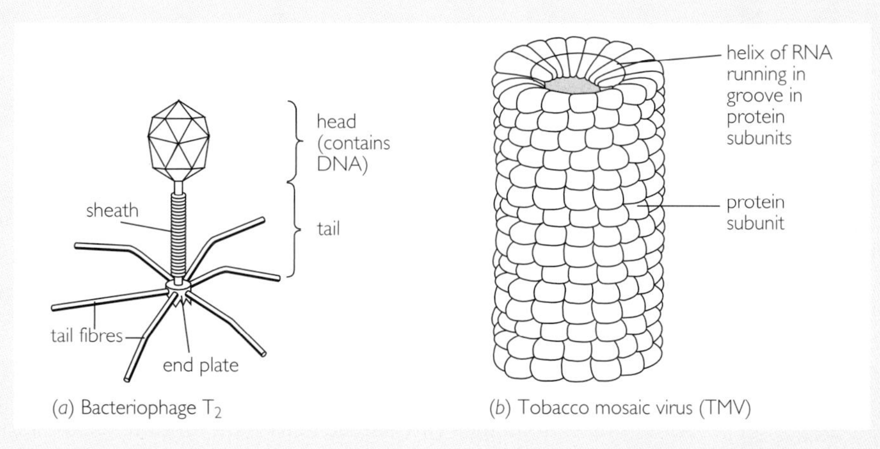

Figure 4.19 Structure of (a) bacteriophage T2; (b) tobacco mosaic virus.

Transport across membranes

QUESTION

Explain the advantages of the division of the cytoplasm into compartments by the membranes of the endoplasmic reticulum in eukaryotic cells.

In order to function efficiently, cells need to be able to take up substances from their immediate environment, to **secrete** useful substances such as **enzymes** and **hormones** manufactured within the cell and to **excrete** the waste products. For aerobic respiration to occur, oxygen and glucose are required by the cell. Carbon dioxide is produced as a waste product and has to be got rid of, or excreted, from the cell.

This movement of molecules into and out of cells involves crossing the cell surface membrane. In eukaryotic cells, the intracellular membranes of the endoplasmic reticulum and other organelles effectively divide up the cytoplasm into compartments, preventing the free movement of molecules. The cell surface membrane and the intracellular membranes are said to be **selectively**, or **partially permeable**. They have a very similar structure, so the ways in which molecules pass across these membranes is also similar. In order to understand the passage of molecules through these membranes, it is necessary to have an understanding of the fluid mosaic model of membrane structure. Due to the hydrophobic nature of the fatty acid chains of the phospholipids, it is difficult for water-soluble molecules to penetrate this barrier and so specific transport systems are required.

Materials enter and leave cells by:

- diffusion
- osmosis
- active transport
- endocytosis and exocytosis.

Diffusion

The rate at which diffusion occurs depends on:

- the concentration gradient: the greater the difference in concentration between the two regions, the greater the rate
- the size of the ions or molecules: the smaller they are, the greater the rate
- the distance over which diffusion occurs: the shorter the distance, the greater the rate.

DEFINITION

Diffusion is the movement of molecules or ions from a region where they are at a high concentration to a region of lower concentration. The difference in concentration is referred to as a **concentration gradient**. There will be a net movement down the concentration gradient until equilibrium is reached, that is when there is a uniform distribution of the ions or molecules. The process is **passive**, as it does not require metabolic energy (Figure 4.20).

Diffusion is affected by the presence of barriers, such as the cell surface membrane and the internal membranes within cells surrounding organelles (Figure 4.20). These membranes are freely permeable to the respiratory gases oxygen and carbon dioxide, which are able to diffuse rapidly in solution depending on the concentration gradients, but are selectively permeable to other molecules. Due to the hydrophobic nature of the membranes, uncharged and lipid-soluble molecules diffuse through more readily than ions and small polar molecules such as glucose and amino acids. The polar molecules are thought to pass through membranes via channels formed by **transport proteins**, also called **channel proteins**.

Some polar molecules can diffuse across membranes more rapidly by combining with special transport proteins. An example of such a mechanism is shown by the movement of glucose molecules into cells (Figure 4.21).

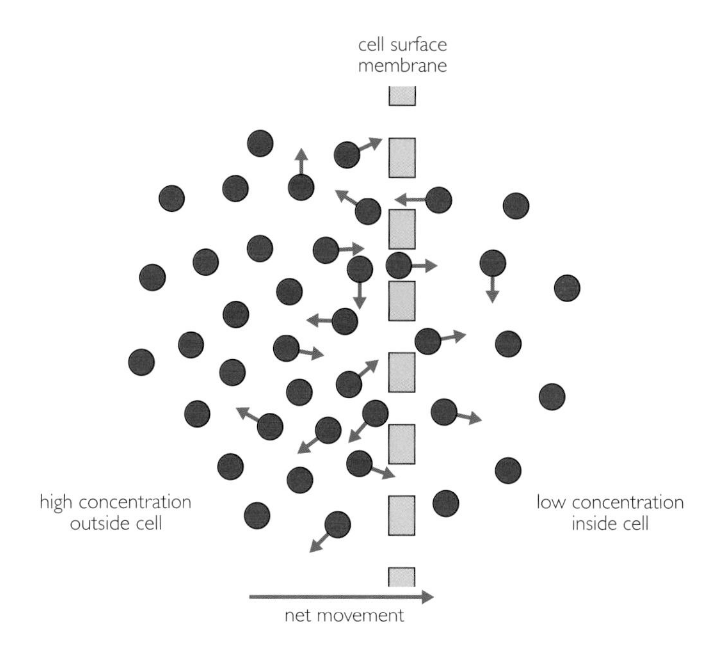

Figure 4.20 Diffusion of molecules occurs down the concentration gradient, from high to low.

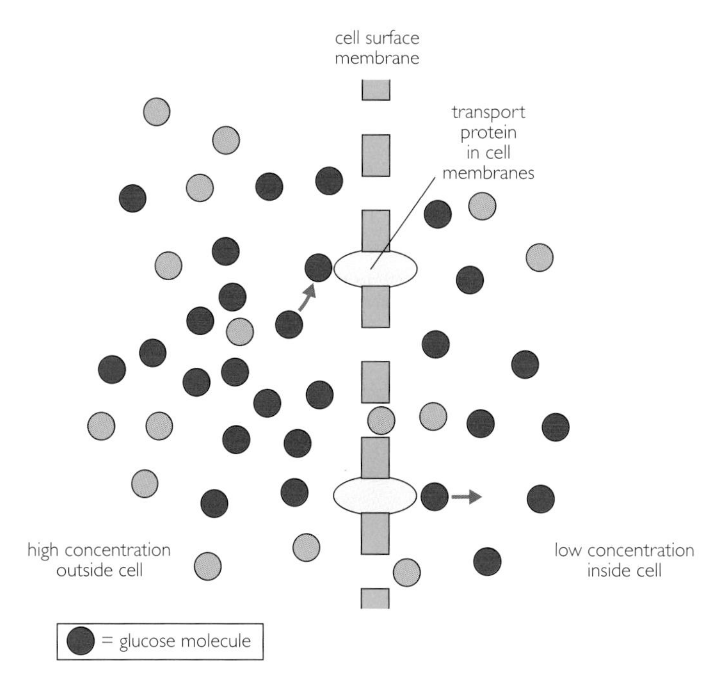

Figure 4.21 Facilitated diffusion is a passive process in which certain molecules bind with membrane proteins to speed up diffusion.

Transport proteins with specific binding sites for glucose are present in the cell surface membrane. Once binding with a glucose molecule has occurred, the protein changes shape and moves the glucose to the other side of the membrane. Once the glucose has become detached, the protein reverts to its original shape and position in the membrane, ready to pick up another glucose molecule. This process is called **facilitated diffusion**. It does not require metabolic energy and it occurs down the concentration gradient until equilibrium is reached.

QUESTION

Compare diffusion with facilitated diffusion.

The diffusion of ions across membranes is affected not only by the concentration gradient, but also by an electrical gradient. The ions will be attracted to areas of opposite charge and will move away from areas of similar charge. Their movement will be governed by a combination of concentration and electrical gradients known as an **electrochemical gradient**. The interior of most cells tends to be negatively charged, favouring the uptake of positively charged ions and repelling negatively charged ones.

Osmosis

Biological membranes are permeable to water, but there is no net movement of water into and out of cells unless **osmosis** occurs, where the movement of the water molecules is linked to the movement and concentrations of solutes on either side of the membrane.

The **water potential** of a solution is the tendency for water molecules to leave or enter that solution by osmosis. Pure water has a water potential of zero. The effect of dissolving a solute in water is to reduce the number of water molecules and lower the water potential; therefore solutions have water potentials with negative values. Water will diffuse from an area of high water potential (less negative) to an area of lower water potential (more negative).

The water potential of a solution is affected by the presence of solute molecules. The dissolved solutes give rise to a **solute potential**; the more dissolved solutes present, the lower (more negative) the solute potential and thus the lower the water potential of the solution. The other factor that affects the water potential is the hydrostatic pressure to which the water is subjected. If the pressure applied to pure water or a solution is increased above atmospheric pressure then the water potential increases.

> **DEFINITION**
>
> **Osmosis** is the diffusion of water molecules from a higher (less negative) water potential to a lower (more negative) water potential through a partially permeable membrane.

Osmosis can be demonstrated using a simple **osmometer**, consisting of a piece of dialysis tubing containing a concentrated solution of sucrose, tied securely at both ends (Figure 4.22). The tubing is immersed in a beaker of water and left for several hours. The dialysis tubing acts as a

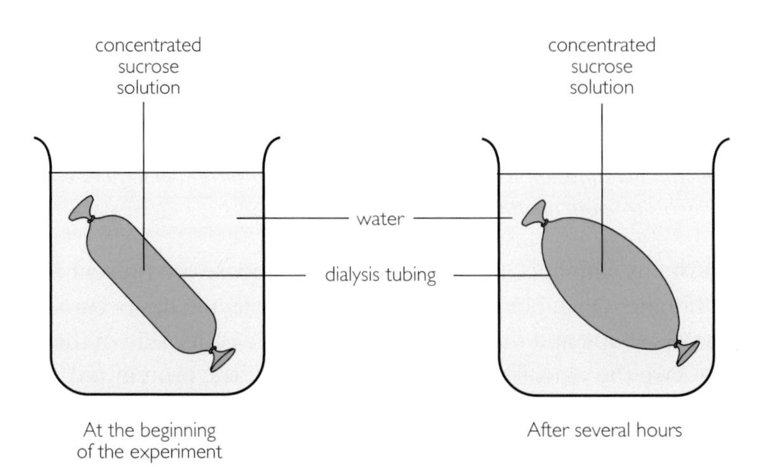

Figure 4.22 Demonstration of the osmotic movement of water through a partially permeable membrane.

partially permeable membrane, allowing water molecules to pass through it but preventing the movement of the sucrose molecules in the opposite direction. When it is examined later, it is seen to have increased in volume due to the entry of water molecules. A concentration gradient exists between the water in the beaker (high concentration of water molecules) and the concentrated solution of sucrose (low concentration of water molecules), so there is a net movement of water molecules into the dialysis tubing bag. Sucrose molecules can diffuse through the dialysis tubing, but because of their size they do so much more slowly than the water molecules.

The topic of water relations of plant and animal cells is discussed in more detail in *Exchange and Transport, Energy and Ecosystems*.

Active transport

Active transport involves the movement of molecules across a membrane up a concentration gradient. Ions are moved up their electrochemical gradients. This movement usually occurs in one direction and requires energy. Most active transport systems are driven by metabolic energy derived from ATP. Active transport allows cells to take up and accumulate ions and molecules necessary for metabolism as well as enabling the waste products to be removed.

The mechanisms used to move the ions or molecules across the membranes are referred to as pumps, the most widespread of which is the **sodium pump**. The majority of animal cells are able actively to pump out sodium ions against the concentration gradient. Usually, at the same time, potassium ions are actively pumped in. This combined mechanism is called a **sodium–potassium pump** and is important in controlling cell volume by removing sodium ions, thereby reducing the tendency of the cell to take up water by osmosis. In addition, the accumulation of potassium ions for use in cell activities, such as protein synthesis, is achieved. The sodium–potassium pump can also be linked to the active uptake of organic molecules, such as glucose and amino acids.

The pump is thought to consist of a protein that spans the cell membrane (Figure 4.23). On the inside of the cell sodium ions bind to special receptor sites on the protein. These ions trigger the phosphorylation of the protein, releasing energy from ATP. The protein changes shape and releases the sodium ions to the outside of the cell. Potassium ions outside the cell bind at another receptor site, causing dephosphorylation of the protein, which then changes back to its original shape, releasing the potassium ions to the inside of the cell.

There is a tendency for the sodium ions to diffuse back into the cell, down their concentration gradient, but this is a slow process because the membrane is less permeable to sodium than to potassium.

Two situations, where glucose is actively transported into cells, can be used to illustrate how the sodium–potassium pump is linked to other transport mechanisms. First, after a meal, there is a high concentration of glucose in the intestine and glucose is absorbed by diffusion down the concentration gradient. This passive process is supplemented by active uptake involving a **glucose transporter protein,** which has binding sites for both glucose molecules and

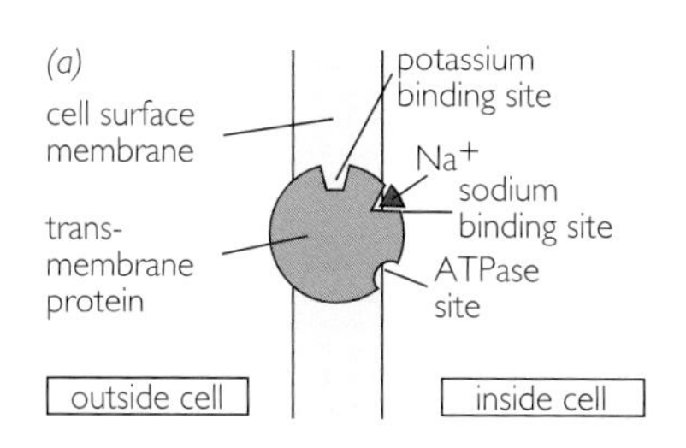

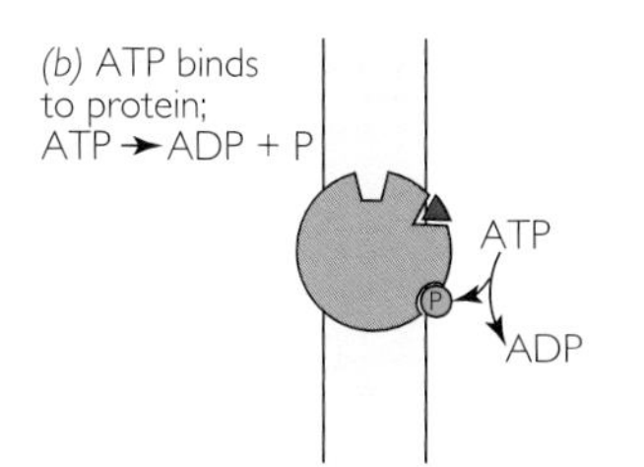

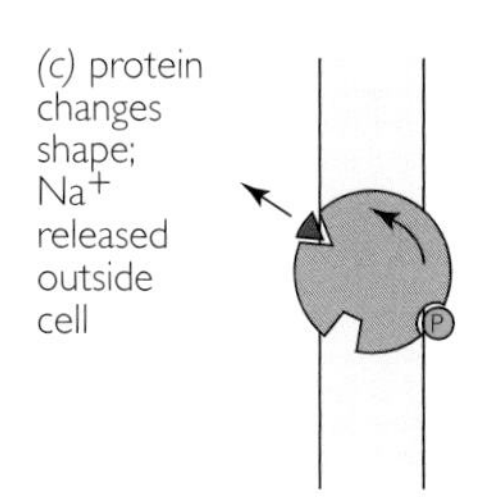

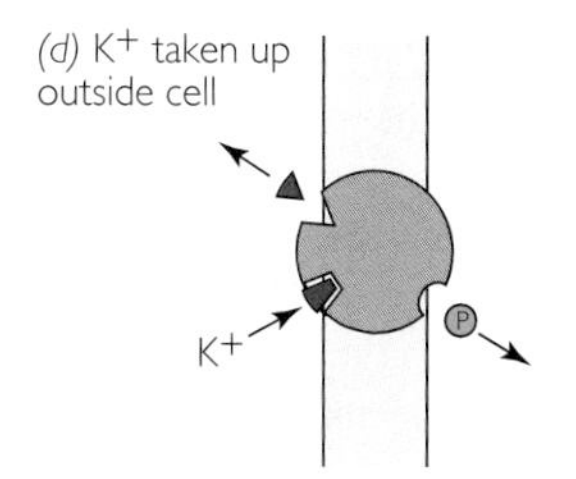

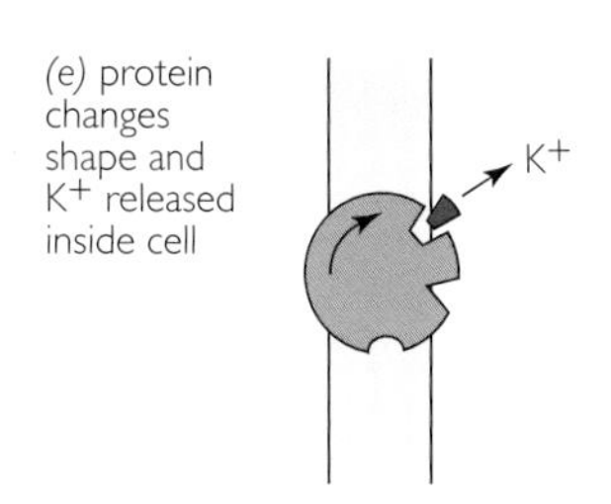

▲ Sodium ion
◆ Potassium ion
Ⓟ Phosphate

Figure 4.23 The sodium–potassium pump is an energy-requiring mechanism in which sodium and potassium ions are exchanged across a cell surface membrane.

sodium ions, present in the membrane of the cells lining the intestine. Second, similar membrane proteins are present in the cells lining the kidney tubules, where glucose is actively taken up against the concentration gradient from the glomerular filtrate. In both situations, the sodium–potassium pump actively transports sodium ions out of the cells against the electrochemical gradient. Glucose molecules and sodium ions bind to the glucose transporter proteins in the cell surface membranes. The sodium ions then diffuse into the cells down their electrochemical gradient, carrying the glucose molecules with them. Once inside the cells, the glucose molecules and sodium ions dissociate from the transporter protein. The glucose concentration of the cells increases, becoming higher than its concentration in the blood, so glucose moves by facilitated diffusion out of the cells into the blood plasma.

Endocytosis and exocytosis

These two processes are involved with the bulk transport of materials through membranes. **Endocytosis** involves the uptake of materials into cells and **exocytosis** is the way in which materials are removed from cells. Both processes depend on the fluidity of the cell surface membrane. The molecules making up the membrane are held together by weak bonds, such as hydrogen and ionic bonds, and hydrophobic interactions, and it has been demonstrated that both the lipid molecules and the proteins can move about in the bilayer.

In endocytosis, there is an invagination of the cell surface membrane to form a vesicle around the material to be taken in, or ingested. The vesicle is pinched off and the cell surface membrane rejoins. The vesicle moves into the cytoplasm where, depending on the circumstances, other vesicles may fuse with it. Substances may be secreted from cells by the reverse of this process, in exocytosis. The vesicles move towards the cell membrane and fuse with it, releasing their contents to the exterior.

> **DEFINITIONS**
>
> **Endocytosis** is the formation of a vesicle or vacuole at the cell surface to enclose extracellular material and take it into the cell.
>
> **Exocytosis** is the fusion of a cytoplasmic vesicle or vacuole with the cell surface membrane so that its contents are removed from the cell.

Endocytosis and exocytosis are involved when solid materials are taken up by cells **(phagocytosis)** and also when the material taken up is a liquid or a suspension **(pinocytosis)**.

Phagocytosis can occur in many situations. It is the mechanism by which protoctists, such as *Amoeba*, ingest their food. The vesicles formed in *Amoeba* are referred to as **food vacuoles**. Certain white blood cells, the **neutrophils** and **monocytes,** are able to recognise foreign bacteria in the blood, engulf them and break them down within a vesicle. The food particles or invading bacteria are taken into the cells by endocytosis and, after digestion, any undigested remains are removed from the cells by exocytosis (Figure 4.24). Once the phagocytic vesicle or food vacuole has been formed, lysosomes, containing digestive enzymes, fuse with it. The contents of the lysosomes are released into the vesicle. It is important to remember that the contents of the vesicles are surrounded by a membrane and so are kept separate from the rest of the cytoplasm. Any products of the digestion of particles inside the vesicles have to diffuse through the membrane in one of the ways already described.

Pinocytosis is also widespread in both the plant and animal kingdoms. It is found in the amoeboid protoctists, white blood cells and cells in the embryo,

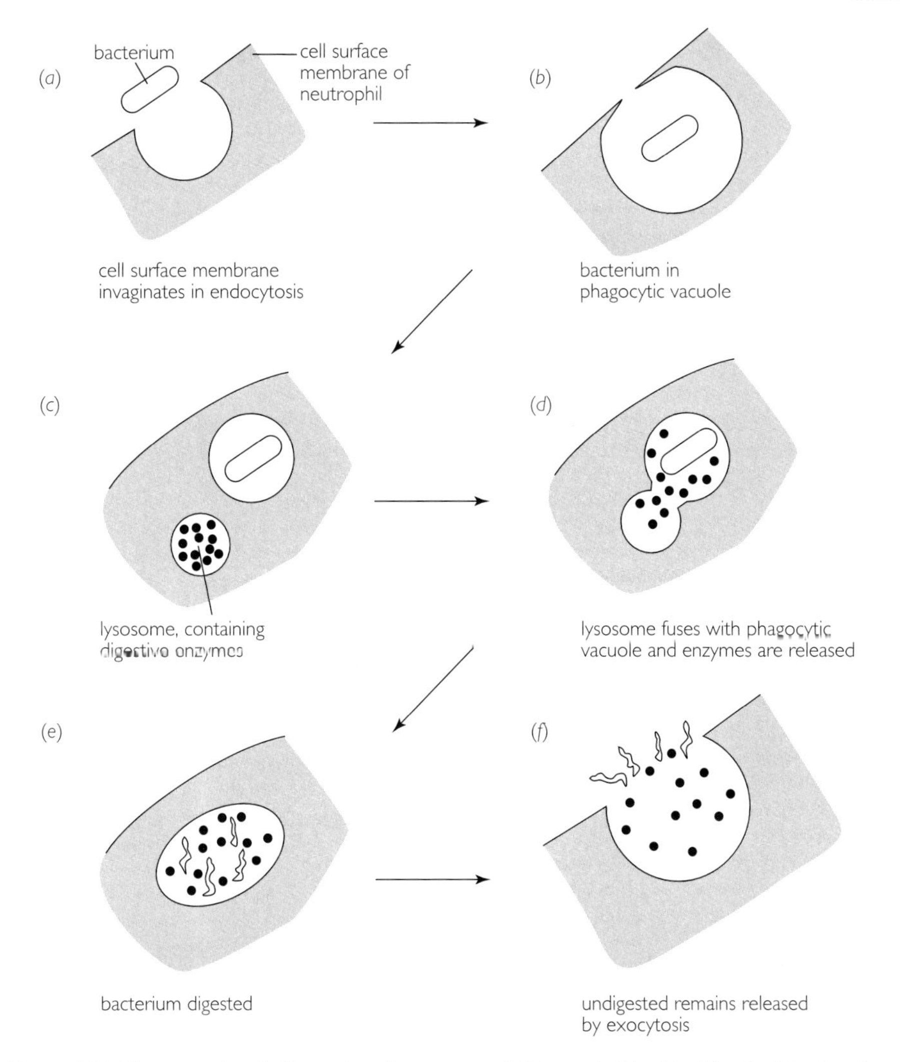

Figure 4.24 Phagocytosis of a bacterium by a neutrophil (amoeboid white blood cell), showing stages of endocytosis, intracellular digestion and exocytosis.

liver and kidneys. The process is essentially similar to phagocytosis, except that liquids are taken up. Sometimes the vesicles formed are extremely small, in which case they are referred to as **micropinocytic vesicles**. These tiny vesicles are usually only detectable on electronmicrographs and have been found in many different types of cells. They have been observed at the base of microvilli on the cells of the epithelium of the small intestine (Figure 4.25), where they are thought to be linked with the uptake of fat droplets.

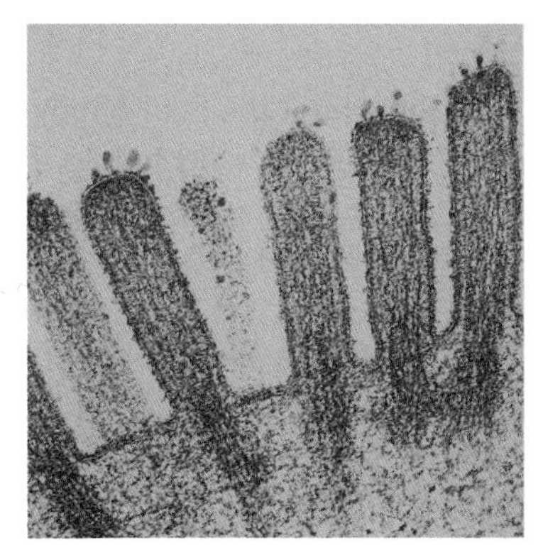

Figure 4.25 Electronmicrograph of microvilli along the edge of cells in the small intestine.

Exocytosis of liquids is important in secretory cells. Enzymes and other products of secretory cells are formed within the endoplasmic reticulum and modified in the Golgi apparatus. These substances are contained in vesicles that are budded off from the flattened cavities of the Golgi body. The vesicles then move through the cytoplasm to the cell surface, where they fuse with the membrane, and exocytosis occurs, releasing the secretions.

PRACTICAL

Setting up and using a light microscope

Introduction

The aim of this practical activity is to familiarise you with the use of a microscope for the observation of prepared slides. Figure 4.26 shows a typical compound light microscope, similar to those in many school and college laboratories. This microscope has an **eyepiece** and three **objective** lenses. The eyepiece usually has a magnifying power of × 10. The objective lenses give magnifications of × 4, × 10 and × 40. The overall magnification is found by multiplying the magnifying power of the lenses used, as shown in Table 4.3 below.

Table 4.3 *Obtaining overall magnification*

Eyepiece magnification	Objective magnification	Overall magnification
× 10	× 4	× 40
× 10	× 10	× 100
× 10	× 40	× 400

The low magnification objective lens (× 4) is used for low-power observation, such as when making a low-power plan; the higher magnifications are used for the observation of cellular details.

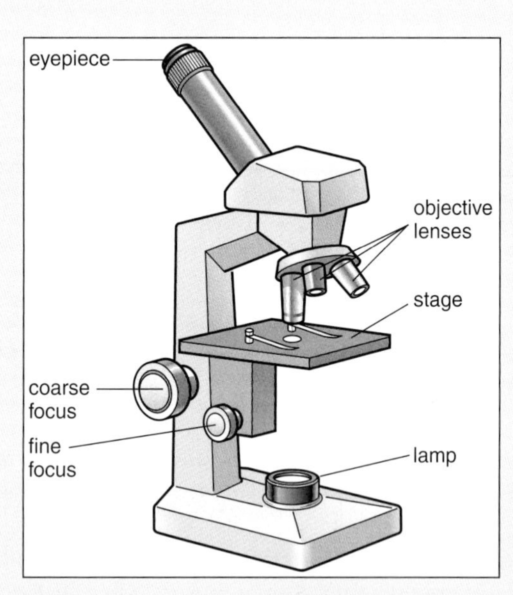

Figure 4.26 A typical compound microscope.

Material

- Microscope
- Prepared slide, such as a transverse section of a leaf, stem or root

Method

1. Turn the low-power objective lens (× 4) into the correct position, directly under the body tube of the microscope. When the objective lens is correctly positioned, a slight 'click' will be heard.

2. Switch on the light or adjust the mirror so that light is reflected up through the tube of the microscope and the field of view is evenly illuminated.

3. Place a prepared slide on the stage so that the specimen is directly under the objective lens.

4. Using the coarse-focus knob, carefully move the objective lens so that it is about 13 mm above the surface of the slide, bringing the specimen into focus.

5. If the microscope is fitted with a condenser, adjust the lighting so that the field of view is uniformly illuminated.

6. When the specimen is in focus with the low-power objective, carefully turn the nosepiece so that the next-power objective (for example, × 10) clicks into position. It should now only be necessary to use the fine-focus knob to bring the specimen into focus.

7. Now turn the nosepiece so that the × 40 objective lens is in position and focus with the fine-focus control.

8. When you have finished viewing the slide, always turn the nosepiece so that the low-power objective (× 4) is in position.

Take care when focusing, particularly with the high-power objectives – do not focus downwards without looking. Always focus with the low-power objective first, before using the high-power objectives.

Measuring the size of an object using an eyepiece graticule

An **eyepiece graticule** consists of a glass or plastic disc with a scale on it. The scale is usually 10 mm long, divided into 100 divisions, referred to as eyepiece units. The eyepiece graticule is inserted into the eyepiece of the microscope, so that when a specimen is observed, the scale is superimposed on the specimen. The eyepiece graticule can be calibrated using a **stage micrometer**. This is a special microscope slide with an accurately ruled scale etched onto it. The scale is usually 1 mm long, divided into 100 units, so that each unit represents 10 μm. The stage micrometer is placed on the stage of the microscope and aligned so that the eyepiece scale and the scale in the stage micrometer are aligned. The total length of the eyepiece scale can then be measured against the scale on the stage micrometer and the value of each eyepiece unit can then be calculated.

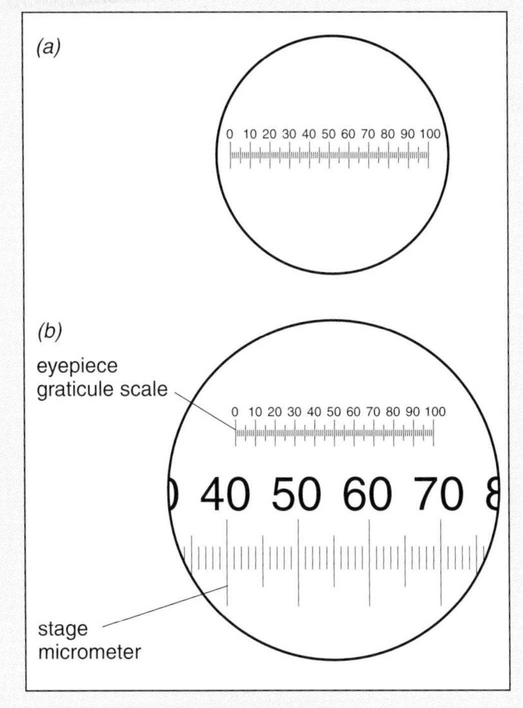

Figure 4.27 (a) An eyepiece graticule; (b) calibration of eyepiece graticule.

As an example, suppose that the 100 units on the eyepiece scale correspond to 30 units on the stage micrometer.

If each unit on the stage micrometer = 10 μm, then 30 units = $30 \times 10 = 300$ μm

Therefore 1 eyepiece unit = $300 \div 100 = 3.0$ μm.

You can then measure the length of an object in the field of view using the eyepiece scale. Multiply the number of eyepiece units by the appropriate factor to find the actual length in micrometres (μm).

This calibration must be carried out separately for each objective lens on a microscope. Eyepiece graticules can be left in position in the eyepiece lenses and, once calibrated, the values for each can be written on a small label and attached to the microscope.

Use a calibrated eyepiece to find the mean lengths and widths of, for example, 10 palisade mesophyll cells in a transverse section of a leaf. Record your results in a suitable table.

Make an accurate labelled drawing to show the structure of a palisade mesophyll cell. Always include a **scale bar** in your drawings to show the actual size of the specimen.

Further details of microscopy, observation and making biological drawings can be found in *Tools, Techniques and Assessment in Biology*.

5 The cell cycle

Stages of the cell cycle

The sequence of events that occurs from the formation of an individual cell until it divides to form daughter cells is called the cell cycle. It is usual to divide the cycle into three stages:

- interphase
- mitosis
- division of cytoplasm (cytokinesis).

QUESTION

Make a list of as many locations in plants and animals where the cell cycle occurs.

Interphase

Observations of nuclei in prepared tissue sections reveal very little taking place during interphase. The nucleus appears as a spherical structure surrounded by a nuclear envelope. The chromatin takes up stains and shows as a granular network in which there are one or more darker staining areas, the **nucleoli**. The nucleoli are not surrounded by a membrane and consist of areas of protein and rRNA. During this stage, which is the longest in the cell cycle, three distinct phases can be distinguished. In the first, shown as G_1 in Figures 5.1 and 5.2, the cell is undergoing a period of rapid growth. New organelles are being synthesised, so the cell requires both structural proteins and enzymes, resulting in a great amount of protein synthesis. The metabolic rate of the cell is high. This phase is followed by the synthesis of new DNA in the nucleus, shown as S on the diagrams. Histones, the proteins to which the DNA is linked, are built up and each chromosome becomes divided into two chromatids. In the third phase, G_2, more cell growth takes place, some of the cell organelles divide, there is an accumulation of energy stores and the chromosomes begin the process of condensation prior to their division.

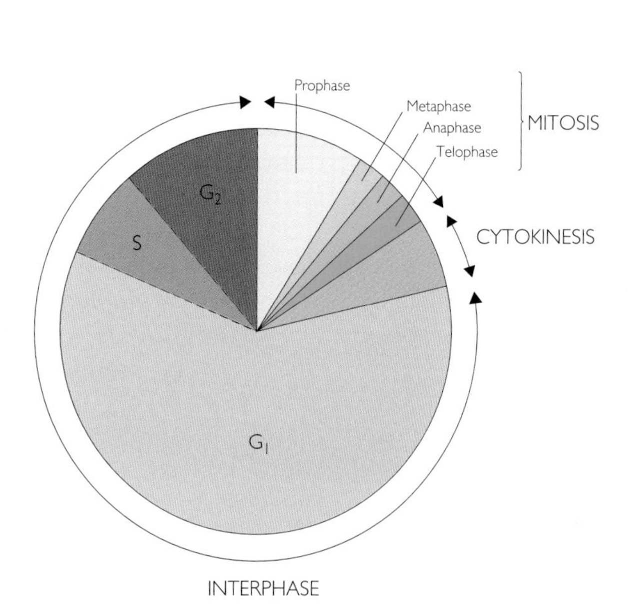

Figure 5.1 The cell cycle, showing the relative durations of the stages in chromosome duplication and separation (mitosis), and cell division (cytokinesis).

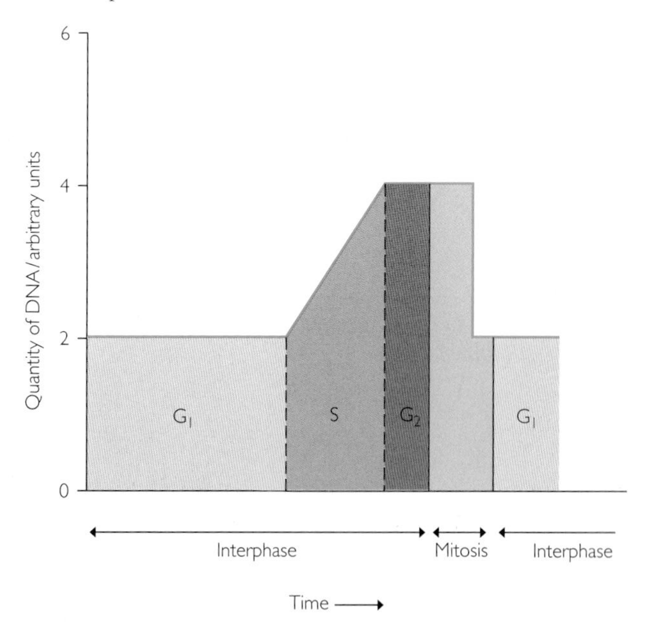

Figure 5.2 Graph to illustrate changes in DNA content of a cell during the cell cycle.

Chromosome number

The number of chromosomes present in the cells of an individual organism is usually constant and is the same for all other individuals of the same species, but the cells of individuals of different species have a different number. For example, onion cells have 16 chromosomes, tomatoes have 24 and humans have 46 (Figure 5.3). This number is referred to as the chromosome number of the species. The chromosomes are seen to be in pairs, that is, there are two identical sets of chromosomes in each cell, so the chromosome number is said to be **diploid**. It is written as 2n, so for the onion 2n = 16, for tomatoes 2n = 24 and for humans 2n = 46. It follows that the gametes normally contain only one of each pair of chromosomes, that is, a single set. They are described as **haploid** and the number is written as n. In the examples we have chosen, the haploid numbers are n = 8 for the onion, n = 12 for tomatoes and n = 23 for humans.

A leaf palisade cell and a liver cell have a diploid chromosome number. These cells have been produced by nuclear division (mitosis) followed by differentiation, during which they have become specialised for their roles.

DEFINITION

The **diploid number** is defined as the number of chromosomes found in the zygote, resulting from the fusion of two **haploid** gametes, and in all the body (or somatic) cells derived from it. A leaf palisade cell and a liver cell have a diploid chromosome number. These cells are produced by nuclear division, followed by growth and differentiation to become specialised for their specific functions.

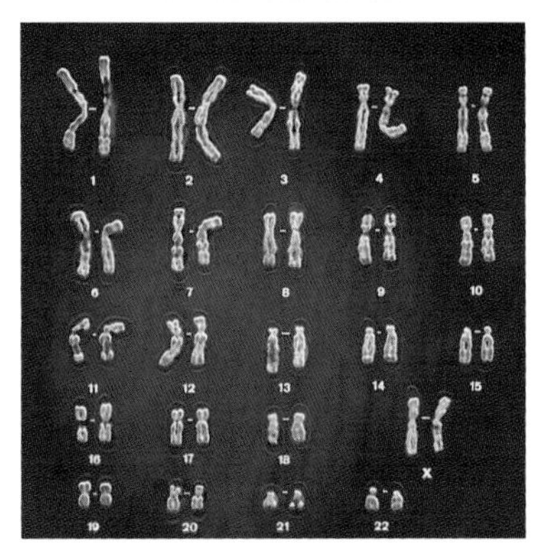

Figure 5.3 Photomicrograph of complete set of 46 human chromosomes from a somatic (body) cell, arranged in 23 similar (homologous) pairs.

Chromosome structure

Chromosomes in eukaryotic cells are made up of:

- DNA (deoxyribonucleic acid)
- proteins
- small amounts of RNA (ribonucleic acid).

Each human chromosome contains one very large molecule of DNA, which, if untangled from its protein, would measure 5 cm in length (Figure 5.4). The total length of DNA in the nucleus of a human cell has been estimated to be about 2.2 m. In order for this enormous length of DNA to fit into the nucleus, a great deal of folding and coiling is involved, especially when cells are undergoing nuclear division. Normally, when we use a microscope to look at the nuclei of cells, we cannot distinguish the chromosomes, but just before nuclear division takes place even more condensation occurs. This results in compact structures that can separate and move during cell division (Figure 5.5). The structure of nucleic acids is described in Chapter 2 and we now need to discuss how the protein and DNA are arranged in chromosomes.

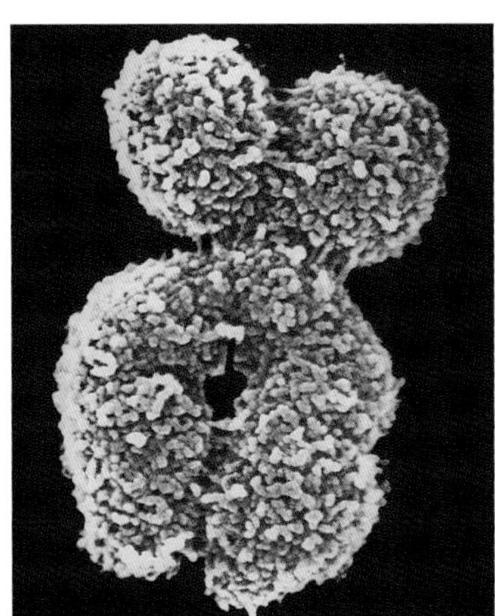

Figure 5.4 False-colour photomicrograph of human chromosome.

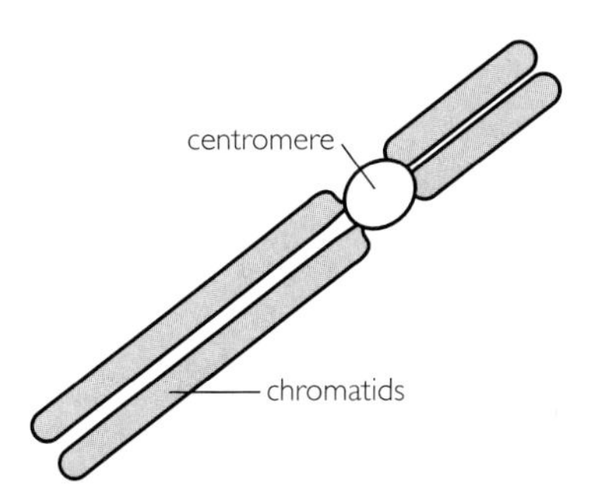

Figure 5.5 Structure of a chromosome showing identical 'sister' chromatids and the centromere, a connecting point essential to chromosome separation during cell division.

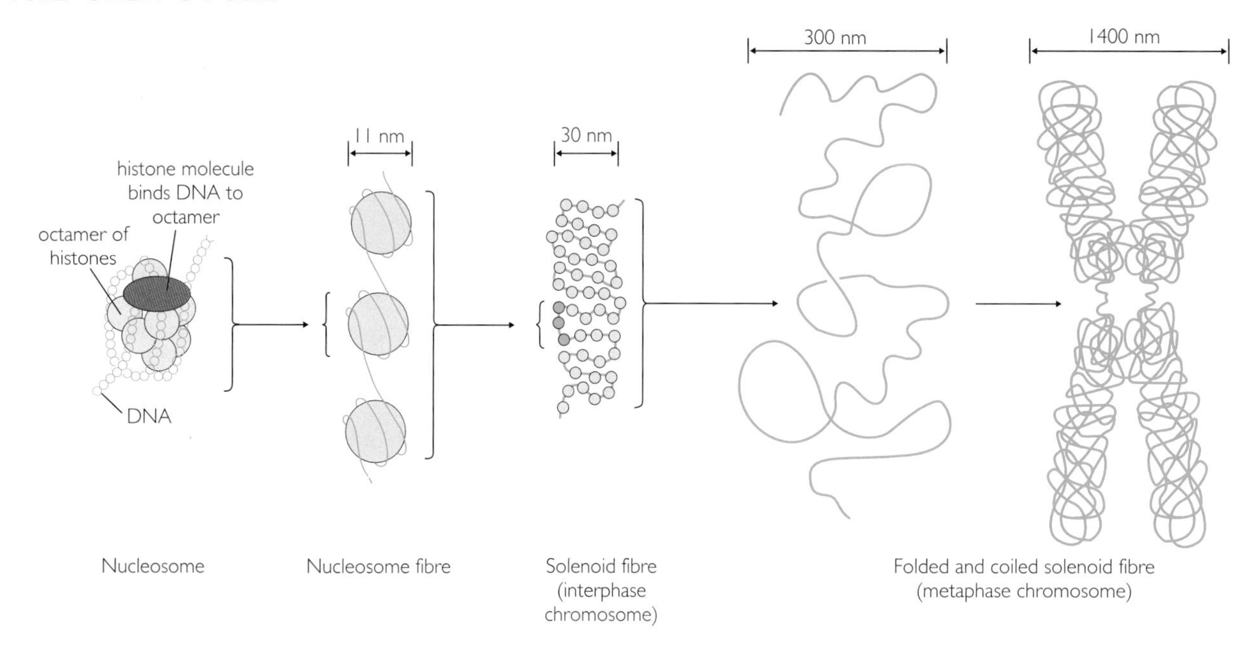

Figure 5.6 Structure of nucleosome and arrangement required to fold and coil 5 cm of DNA into a single chromosome less than 1.5 μm wide.

The protein component of a chromosome consists of:

- histones
- scaffold protein
- polymerases.

The **histone** proteins have large numbers of positively charged amino groups and are referred to as basic proteins. The phosphate groups of the DNA are negatively charged and bind to the amino groups of the histones, forming a stable structure. The complex formed between the DNA and the histones is called chromatin, which exists as fibres 11 nm wide. In prepared slides of plant and animal tissues the chromatin takes up stains, enabling the nuclei of cells to become visible. Unless nuclear division is occurring, individual chromosomes are not visible under the light microscope, but electron microscopy shows the chromatin as having a beaded appearance due to the presence of nucleosomes. Each nucleosome consists of a group of eight histone molecules (called an octamer) around which is wrapped a length of the DNA (Figure 5.6). There are 146 base pairs in the DNA around the nucleosome and this length of the double helix appears to be held in place by another histone molecule attached to the outside. The nucleosome is considered to be the basic unit in the structure of chromatin.

The nucleosome fibre is tightly coiled and this in its turn is thought to be coiled and looped around non-histone proteins collectively referred to as **scaffold** protein. Precise details of this further coiling and folding are not known.

The DNA content of prokaryotic cells is associated with much less protein than that of eukaryotic cells. The double helix of DNA is twisted and sealed, forming a circular shape.

The polymerases are enzymes involved with **transcription** of the genetic information during protein synthesis and with **replication** of the DNA prior to division of the chromosomes.

When it was understood that nucleic acid rather than protein was the molecule of inheritance there was great interest in working out its composition and three-dimensional structure. James Watson and Francis Crick, working in Cambridge in 1953, were the first to show how the polynucleotide chains were arranged in the double helix of DNA. Much of the evidence was provided by Maurice Wilkins and Rosalind Franklin, of King's College, London. They passed X-rays through crystalline DNA and looked at the way in which the X-rays are scattered by the atoms in the molecule (Figure 5.7). These X-ray diffraction patterns were interpreted by Watson and Crick, who then built a model to represent DNA.

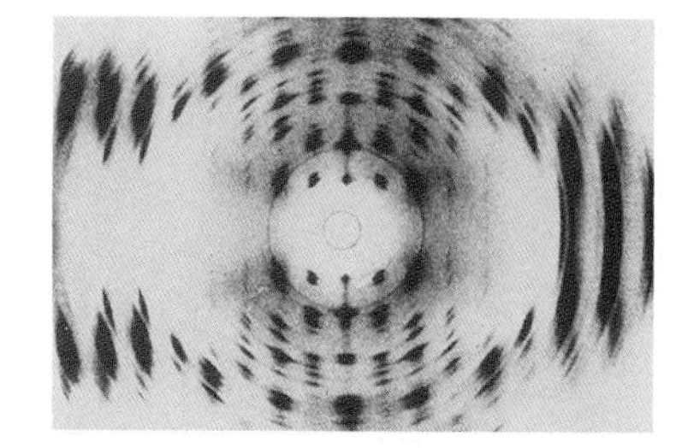

Figure 5.7 X-ray diffraction photograph (top) of DNA taken by Maurice Wilkins, which helped Francis Crick and James Watson deduce the double-helix structure of DNA and model it (above).

The structure of the DNA molecule and its replication are described in Chapter 2, but it is relevant here to summarise its main features in the light of its suitability for carrying genetic information:

- it is a double helix composed of two polynucleotide chains
- each polynucleotide chain has a sugar phosphate backbone on the outside with the bases on the inside
- hydrogen bonding occurs between specific bases on opposite chains forming base pairs, holding the two chains together
- adenine (a purine) only pairs with thymine (a pyrimidine)
- guanine (a purine) only pairs with cytosine (a pyrimidine)
- the two chains are complementary; the sequence of nucleotides in one chain determines the sequence of nucleotides in the other chain
- the chains are **anti-parallel**, that is, the 3′ end of one chain lies next to the 5′ end of the other
- it is capable of replication.

Mitosis

Interphase is followed by mitosis, the nuclear division that involves the separation of sister chromatids and their distribution into the daughter nuclei. The events that take place during this separation are continuous, but for descriptive purposes it has been convenient to recognise four main stages: **prophase**, **metaphase**, **anaphase** and **telophase**. Overall, the process involves condensation of the chromosomes and their precise arrangement in the cell so that, when the chromatids are pulled apart, they are divided exactly into two identical groups (see *Practical: Preparation of a root tip squash*).

Prophase: In prophase, the chromosomes first appear as long tangled threads that gradually become shorter and thicker, due to spiralisation. Eventually they are seen to consist of two chromatids, which are held together at an unstained region called the centromere. The nucleoli get smaller and gradually disappear. In most animal and some plant cells, the **centrioles**, situated just outside the nuclear envelope in the cytoplasm, move to opposite ends (poles) of the cell. **Microtubules** radiate out from the centrioles, forming an aster. When all these events have taken place, the nuclear envelope breaks up and microtubules are organised to form a barrel-shaped structure, called a **spindle** (Fig. 5.8), across the central part of the cell, between the centrioles. The microtubules involved in the spindle are referred to as **spindle fibres** (see Figures 5.9a, 5.10a).

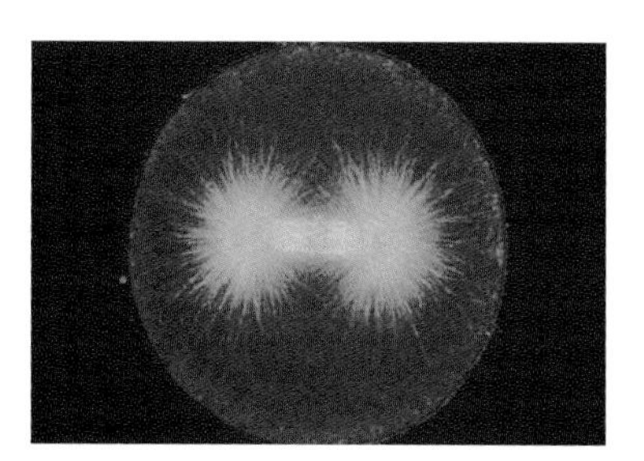

Figure 5.8 Photomicrograph of cell division spindle formation in sea urchin eggs.

Metaphase: During metaphase, the chromosomes become attached to the spindle fibres by their centromeres. They move up and down so that they are aligned along the central part, or equator, of the spindle. They are arranged at right angles to the axis of the spindle and the sister chromatids of each chromosome are easily distinguished (Figures 5.9b, 5.10b.).

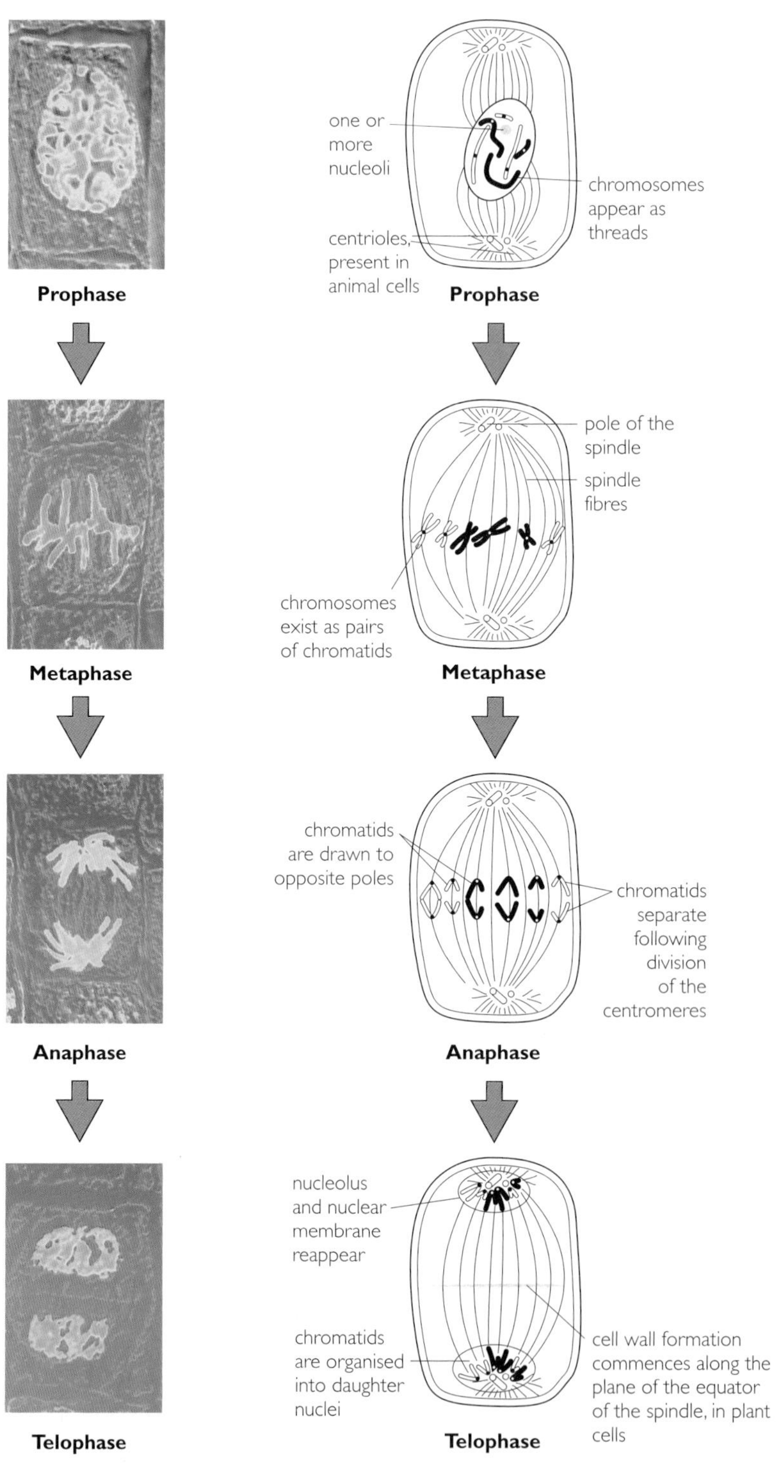

Figure 5.9 Photomicrographs of mitosis in Hyacinth *root tip cells, showing the four main phases of division: (a) prophase; (b) metaphase; (c) anaphase; (d) telophase.*

Figure 5.10 Diagrammatic representations of the four phases of mitosis: (a) prophase; (b) metaphase; (c) anaphase; (d) telophase.

Anaphase: The centromeres divide and the spindle fibres shorten, pulling the centromeres to opposite poles. This is anaphase and results in the separation of the chromatids. Once separated, the chromatids are now referred to as daughter chromosomes (Figures 5.9c, 5.10c).

Telophase: When they reach the poles of the cell, at telophase, the chromosomes begin to lengthen, uncoiling and losing their visibility. A nuclear envelope forms around each group of daughter chromosomes, nucleoli reappear and the division of the nucleus is completed (Figures 5.9d, 5.10d).

The significance of mitosis

The information held on the chromosomes in the nucleus is used to manufacture cell components and to control the metabolic activities of cells, through the production of enzymes and hormones. It is essential that this information is passed on to new cells produced within an organism. The process of mitosis ensures that this happens, as two genetically identical daughter nuclei are formed, each containing the same number of chromosomes as the parent cell.

> **DEFINITION**
>
> **Mitosis** is the division of a nucleus to produce two daughter nuclei, each containing the same genetic information as the parent nucleus. Each daughter nucleus contains the same number of chromosomes and these chromosomes are genetically identical to the chromosomes in the parent nucleus.

Mitosis results in an increase in the numbers of cells and is associated with the growth of an individual organism, or of a colony if the organism is unicellular. In addition, it is found where repair and replacement of tissues occurs, such as in the formation of callus tissue where wounding has occurred in flowering plants, and in the replacement of red blood cells in vertebrates. In the life cycles of prokaryotes, protoctists and fungi, it is associated with asexual reproduction.

The essential features of mitosis are that:

- the chromosome number is maintained
- there is no change in the genetic material.

Cytokinesis

Cytokinesis follows telophase. The other cell organelles, such as ribosomes and mitochondria, become evenly distributed around each nucleus and the process of division of the cytoplasm begins. In animal cells, the process is sometimes referred to as **cleavage**, since a furrow develops by intucking of the cell surface membrane. It is thought that protein fibres in the cytoplasm, called **microfilaments**, are involved. The furrow becomes deeper until eventually the two cells separate.

In plant cells, the spindle fibres in the equator region of the cell do not disappear but move outwards, forming a structure known as a **phragmoplast.** Many cell organelles congregate in this area and a number of fluid-filled vesicles are budded off from the Golgi apparatus. These vesicles contain material that is needed to build a middle lamella and a new cellulose cell wall. The vesicles join up together to form the **cell plate**, which grows across the middle, eventually separating the two daughter cells (Figure 5.11). The membranes of the vesicles contribute to the new cell surface membranes. In certain regions, the vesicles do not fuse, leaving a cytoplasmic connection called a **plasmodesma** (plural: **plasmodesmata**) between the two daughter cells.

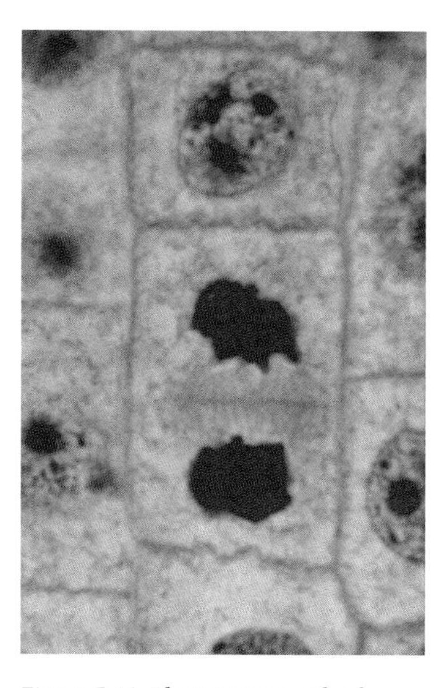

Figure 5.11 Photomicrograph of cleavage and cell-plate formation in dividing onion (Allium) *root tip cells.*

Natural and artificial cloning in plants and animals

Offspring produced by asexual reproduction are genetically identical to the parent organism and are described as a **clone**. Yeast cells produced by budding, which involves mitosis, are natural clones of the parent yeast cell. All members of a clone are genetically identical.

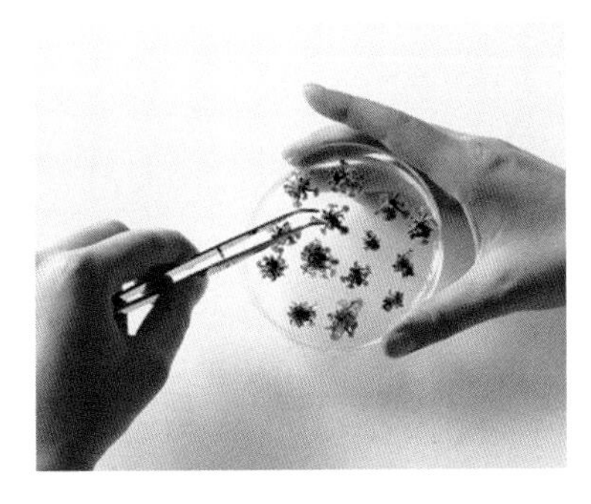

Figure 5.12 Genetically identical plantlets (clones) produced by tissue culture.

Body cells from multicellular organisms, such as vertebrates or flowering plants, can be grown in tissue culture and induced to divide. In the technique of micropropagation, large numbers of genetically identical plants of the same variety can be produced very quickly. Tissue from the apical meristem of a desired plant is cultured on media containing nutrients and growth substances under carefully controlled, sterile conditions. The growth substances promote root and shoot development. When the plants have developed sufficiently, they can be placed in glasshouses to harden off, but need to be kept humid until the cuticle on the surface of the leaves and stems has developed fully.

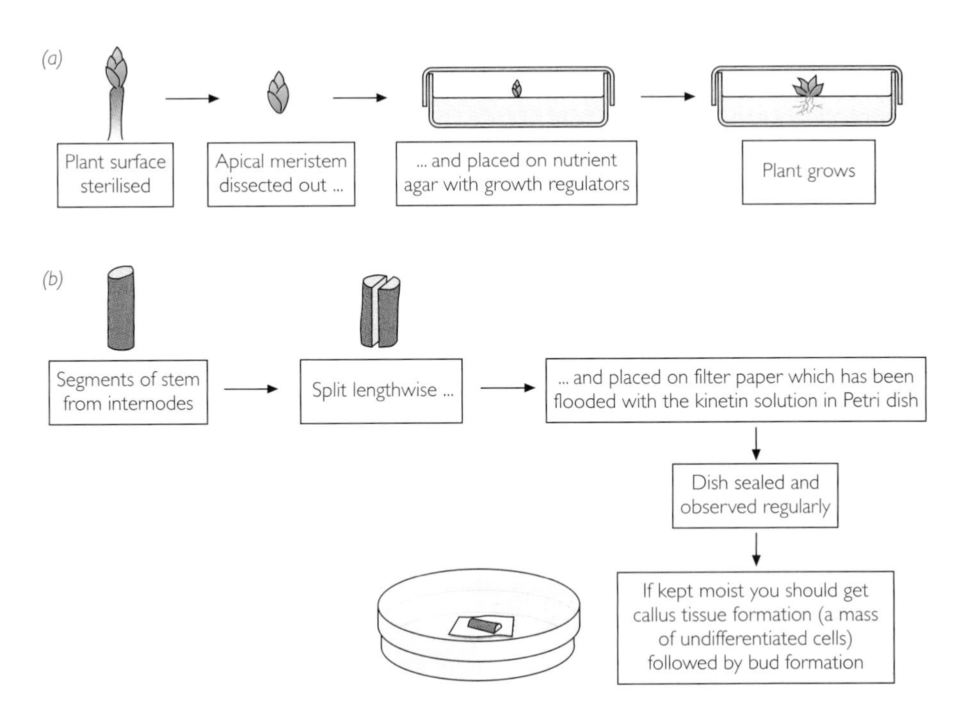

Figure 5.13 Tissue culture is used in commercial crop propagation to produce large numbers of genetically identical plants; (a) and (b) above show two different ways of carrying out micropropagation.

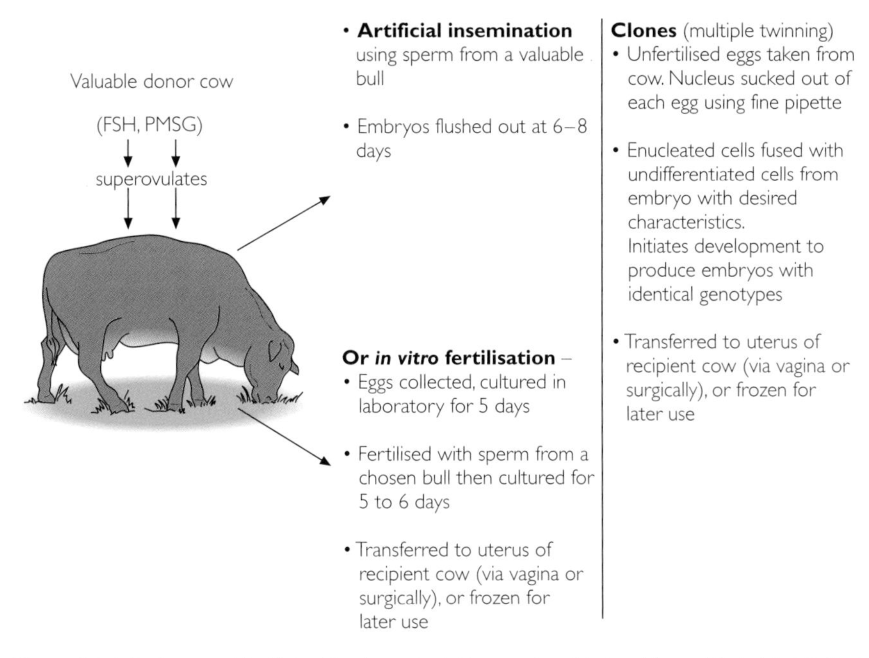

Figure 5.14 Embryos with disirable characteristics can be obtained by artificial insemination or in vitro *fertilisation. These embryos can be used for cloning.*

Artificial cloning has been successfully carried out in animals. It is possible to remove the nuclei from unfertilised eggs, using a fine pipette. These enucleated eggs are then fused with undifferentiated cells from an embryo with desired characteristics. The resulting embryos have identical genotypes. This technique is possible in cattle, where cows can be stimulated to superovulate, resulting in large numbers of eggs being released at the same time. The cloned eggs can then be introduced into recipient cows, known as surrogate mothers. These cows are non-pedigree surrogates and the cloned eggs have to be introduced at the correct stage of the cows' oestrous cycle. The technique enables large numbers of offspring with desirable characteristics to be produced, compared with one calf per year by conventional breeding.

Cloning from an adult animal has been achieved in the production of 'Dolly' the sheep. The techniques are slightly different from those described above, but the principle is the same, resulting in a genetically identical individual. There is currently much debate about the value of cloning, particularly from adult animals. On one hand, cloning reduces genetic diversity if large numbers of identical offspring are produced. On the other hand, it can be a useful technique in the conservation of desirable characteristics and, in the long run, may serve to increase genetic diversity.

Practical

Preparation of a root tip squash

Introduction

Garlic root tips provide a reliable source of actively dividing cells to demonstrate mitosis. Individual garlic cloves, supported over water using cocktail sticks, will produce numerous roots within 4 to 5 days. The demonstration of chromosomes is by the Feulgen reaction. Acid hydrolysis of DNA results in the formation of aldehydes, which react with Schiff's reagent to produce a bright red-purple colour. This method is specific for DNA.

Materials

- Actively growing garlic roots
- Fixative (99 cm^3 70 per cent aqueous industrial methylated spirits plus 1 cm^3 glacial ethanoic acid)
- 1.0 molar hydrochloric acid
- 70 per cent aqueous glycerol
- Distilled water
- Schiff's reagent
- Water bath at 60 °C
- Test tubes
- Microscope slides and coverslips
- Scalpels
- Mounted needles
- Blotting paper
- Microscope

FLAMMABLE
methylated spirits

CORROSIVE
ethanoic acid
hydrochloric acid
Schiff's reagent

Method

1 Cut off the end 1 cm of a root and fix in a mixture of 99 cm^3 70 per cent (aqueous) industrial methylated spirit, plus 1 cm^3 glacial ethanoic acid for at least 2 hours.

2 Treat the root tips in 1 molar hydrochloric acid at 60 °C for 6 to 7 minutes. This is the critical part of the method; if treated for longer the staining reaction becomes weaker.

3 Remove the acid and rinse thoroughly in distilled water.

4 Add Schiff's reagent and leave for 1 hour.

5 Rinse again in distilled water.

6 Place one root tip on a microscope slide, cut off and discard all but the darkly stained tip. Add a couple of drops of 70 per cent aqueous glycerol and, using mounted needles, break up the root tip.

7 Carefully apply a coverslip and place the slide between several sheets of blotting paper. Squash gently. The cells should now be sufficiently well spread out.

8 Examine the preparation using a microscope, first with low magnification, then high. Look carefully for cells with visible chromosomes.

Results and discussion

1 Make labelled drawings of representative stages of mitosis.

2 Count the total number of cells visible in the field of view. Then, count the total number of cells that show stages of mitosis and express this as a percentage of the total number of cells present in the field of view. This percentage is known as the **mitotic index** and is a measure of the proportion of the time spent in mitotic division.

Appendix: Physical Science Background

All Advanced Subsidiary (AS) and Advanced GCE Biology and Human Biology specifications (syllabuses) assume that you have some background knowledge of physics, chemistry and mathematics. You may already have met some of these topics at Key Stage 4 (or earlier) and others will be introduced during your AS or GCE A studies. Alternatively, you may not have studied science before taking your GCE A course, in which case this section introduces you to some of the background in physical science.

To answer questions set in the Unit Tests, you should have some knowledge and understanding of the following topics.

- ☐ atoms
- ☐ electrons
- ☐ ions
- ☐ elements
- ☐ electrovalent bond
- ☐ covalent bond
- ☐ molecules
- ☐ hydrogen bond
- ☐ latent heat
- ☐ solubility
- ☐ acidity and the pH scale
- ☐ buffering
- ☐ formulae in chemical equations
- ☐ oxidation and reduction
- ☐ condensation
- ☐ hydrolysis
- ☐ energy changes
- ☐ the electromagnetic spectrum
- ☐ pressure
- ☐ partial pressure
- ☐ diffusion

Atoms, electrons, ions and elements

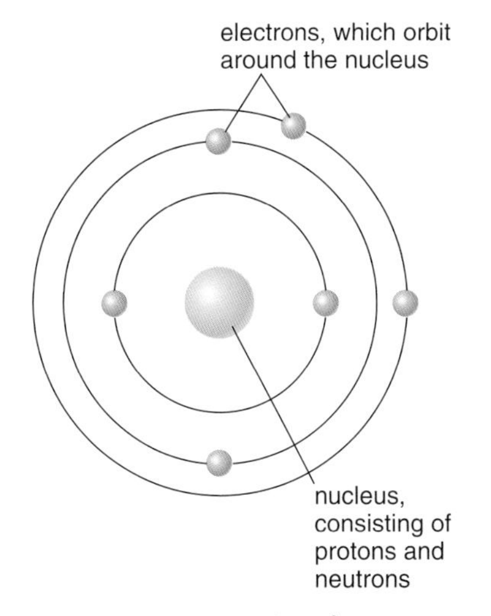

Figure A.1 Representation of an atom. The nucleus consists of protons and neutrons, surrounded by a cloud of orbiting electrons. Protons have a positive charge, electrons have a negative charge, neutrons have no charge.

An **atom** consists of a nucleus, containing **protons** and **neutrons**, surrounded by a cloud of **electrons**. Each proton has a positive (+) charge and the number of these protons (referred to as the **atomic number**) determines the nature of the atom. Neutrons have a similar mass to protons, but they do not have a charge. The atomic mass of an atom is equal to the combined mass of the protons and neutrons. Atoms that occur naturally on the Earth contain from 1 to 92 protons. As examples, a hydrogen atom has just one proton in its nucleus and an oxygen atom has eight. The positive charges in the nucleus of an atom are balanced by negatively (–) charged electrons which orbit around the nucleus. The positive charge of one proton is exactly balanced by the negative charge of one electron. Electrons have very little mass (only about 1/1840 of the mass of a proton).

Sometimes, one or more electrons may be lost by an atom, so that the number of positive charges in the nucleus is no longer balanced by the number of negative charges due to the electrons. An atom in which the number of electrons does not equal the number of protons is known as an ion. Ions have an overall charge: if electrons are lost, the ion will have a positive (+) charge; if electrons are gained, the ion will have a negative (–) charge. As examples, an atom of potassium (chemical symbol K) that has lost an electron becomes a positively charged potassium ion (written K^+); a chlorine atom (symbol Cl) that gains an electron becomes a chloride ion (Cl^-). Since an electron has a negative charge, it is often represented as e^-.

We can write equations to show these changes, as follows.

$K \rightarrow K^+ + e^-$ (that is, an atom of potassium becomes a potassium ion by *losing* an electron)

and

$Cl + e^- \rightarrow Cl^-$ (that is, a chlorine atom becomes a chloride ion by *gaining* an electron)

Calcium (chemical symbol Ca) and magnesium (symbol Mg) both form ions by losing *two* electrons, so a calcium ion and a magnesium ion both have two positive changes (written as Ca^{2+} and Mg^{2+}). The term **cation** is used for an ion with a positive charge; an **anion** is an ion with a negative charge.

There are 92 naturally-occurring types of atoms, each of which is referred to as an **element**. Each element has a characteristic number of protons in its nucleus, which gives the element its atomic number; for example, an atom of calcium, atomic number 20, has 20 protons in its nucleus.

Table A.1 *Some familiar elements with their chemical symbols and atomic numbers*

Element	Symbol	Atomic number
Calcium	Ca	20
Carbon	C	6
Chlorine	Cl	17
Hydrogen	H	1
Iron	Fe	26
Magnesium	Mg	12
Nitrogen	N	7
Oxygen	O	8
Phosphorus	P	15
Sodium	Na	11
Sulphur	S	16
Zinc	Zn	30

Electrovalent and covalent bonds

Relatively few substances occur naturally in the form of a pure element; most atoms are combined to form a stable structure referred to as a **molecule**. Atoms join together to form molecules by:

- gaining one or more electrons
- losing one or more electrons
- sharing electrons with another atom.

We have already noted that, when an atom loses an electron, it become a positive ion. An atom that gains an electron becomes a negative ion. Opposite charges, that is, negative and positive, attract each other so that a positively charged ion and a negatively charged ion will be held together by the attraction of opposite charges, forming an **electrovalent bond** (or ionic bond) between them. A different kind of bond results when two atoms *share* one or more pairs of electrons, thus forming a **covalent bond**.

> **DEFINITION**
> A chemical bond is a force which holds two atoms together. This force results from the attraction of opposite charges, referred to as an **electrovalent bond**, or from the sharing of one or more pairs of electrons, as in a **covalent bond**.

A crystal of sodium chloride, or common table salt (chemical formula NaCl), consists of a regular arrangement of sodium ions (Na^+) and chloride ions (Cl^-) held together by **electrovalent bonds**. The positive and negative charges of the ions attract each other and the sodium and chloride ions cluster together to form a crystal of salt, which is typically cube-shaped.

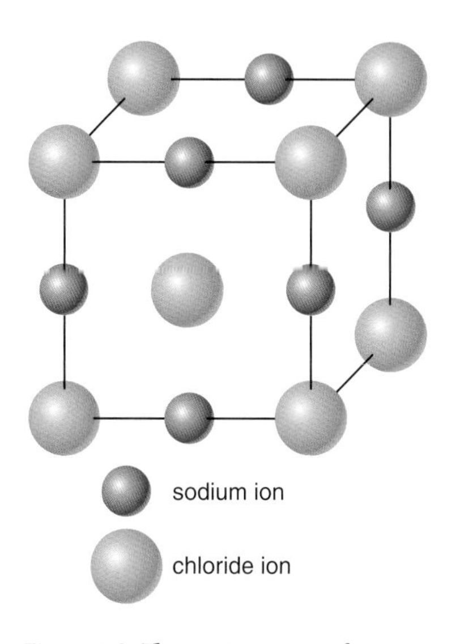

Figure A.2 The arrangement of sodium ions (Na+) and chloride ions (Cl–) in part of a crystal of sodium chloride.

We know that salt dissolves in water; when this happens, water molecules break up the forces holding the ions together in their crystal structure and the salt dissolves to form a solution containing a mixture of free sodium ions (Na^+) and free chloride ions (Cl^-). The physical and chemical properties of water are very important in biology, this topic will be considered below in the section *Properties of water*.

A **covalent bond** is formed when two atoms share one or more pairs of electrons. As an example, an atom of hydrogen (H) has just one electron which orbits the nucleus. A molecule of hydrogen gas (H_2) is formed when two atoms of hydrogen share their electrons, which orbit both nuclei. The two hydrogen atoms form a stable molecule of hydrogen, the bond between the two atoms being an example of a covalent bond. Another example of a molecule formed by two atoms joined together is oxygen gas (O_2), which consists of two atoms of oxygen, joined in this case by the sharing of two pairs of electrons, known as a **double bond**.

> **DEFINITION**
> A **covalent bond** is formed when one atom shares one or more pairs of electrons with another atom. A covalent bond is stronger than an electrovalent bond.

Molecules often consist of more than just two atoms and one reason why larger molecules form is that an atom may be able to share electrons with more than one other atom. All **organic** substances contain carbon atoms and each carbon atom can form four covalent bonds. One simple example of this is the way in which carbon forms four single covalent bonds with four atoms of hydrogen to produce a molecule of methane (CH_4). The four hydrogen atoms are arranged equally around the central carbon atom, forming a pyramid-shaped tetrahedron. A molecule of methane may be represented in different ways, as shown in Figure A.3.

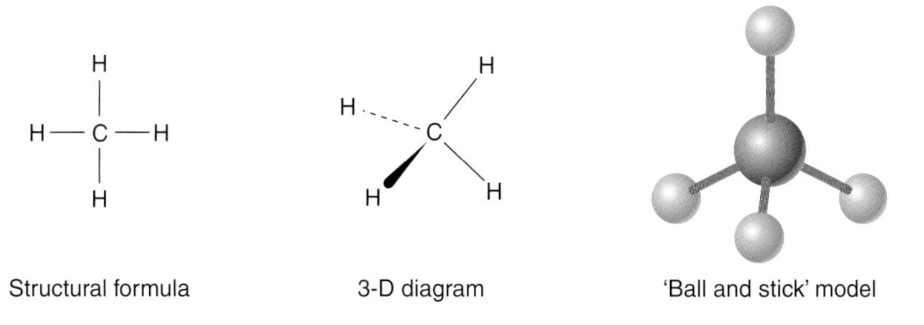

Figure A.3 Three different ways of representing a molecule of methane (CH_4).

Molecules in living organisms

Of the 92 different kinds of atom which exist naturally, only 11 are commonly found in living organisms and the majority of these are carbon (C), oxygen (O), hydrogen (H) or nitrogen (N). Carbon and hydrogen are present in all **organic** substances; carbon forms four covalent bonds and hydrogen forms one. Oxygen forms two covalent bonds and nitrogen three. Carbon, hydrogen, oxygen and nitrogen can join in different combinations to form an almost infinite number of different molecules, including amino acids, carbohydrates, lipids and nucleic acids, all of which are important in living organisms.

To take an example, let us look at the way in which the atoms of carbon, hydrogen and oxygen join together to form a molecule of glucose. The molecular formula of glucose is $C_6H_{12}O_6$. This tells us that one molecule of glucose contains six atoms of carbon, twelve atoms of hydrogen and six atoms of oxygen, but it does not tell us how the atoms are joined together to form the molecule. Look at Figure A.4, which shows a diagram of a molecule of glucose. Identify the carbon, hydrogen and oxygen atoms and note that each carbon atoms has four bonds, each hydrogen atom has one bond and each oxygen atom has two bonds, in accordance with the principles for the formation of covalent bonds.

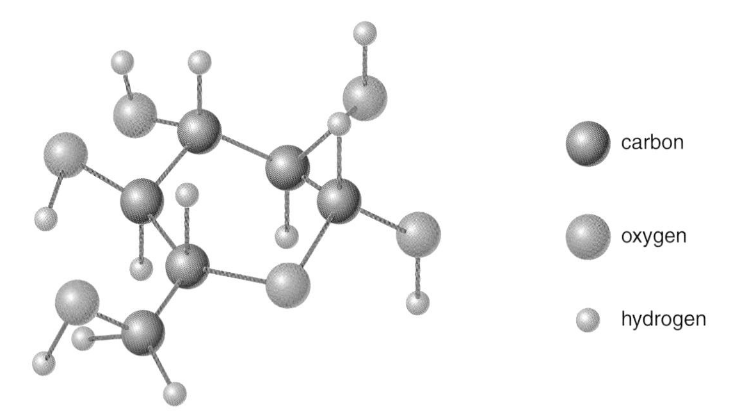

Figure A.4 A 'ball and stick' molecule of glucose ($C_6H_{12}O_6$).

Properties of water: hydrogen bonds, latent heat and solubility

Water has the chemical formula H_2O and consists of an atom of oxygen covalently bonded to two hydrogen atoms. The oxygen atom is surrounded by a stable arrangement of eight negatively charged electrons; however, more of the electrons are associated with the oxygen nucleus than with either of the hydrogen nuclei. This means that the oxygen atom has an overall partial negative charge (often written as δ^-), which is much less than the negative charge of an ion. The electron cloud surrounding the oxygen atom is denser than that around the hydrogen atoms, and this creates partially negative and partially positive charges on the ends of a water molecule. A water molecule is described as being **polar** because of these charged ends. The partially positive and negative charges attract each other and this feature of water molecules gives them a number of important properties. A molecule of water is illustrated in Figure A.5. Notice that there is a bond angle, of about 105°, between the two covalent oxygen–hydrogen bonds.

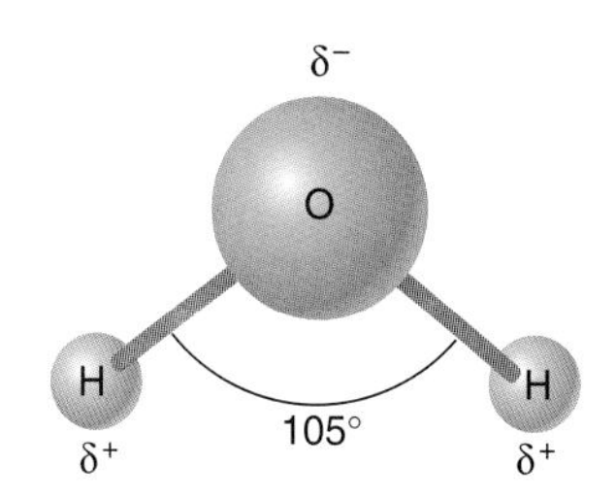

Figure A.5 A model of a water molecule. The oxygen atom has a partial negative charge (δ^-) and the hydrogen atoms have a partial positive charge (δ^+). This makes the water molecule ***polar****.*

The weak forces of attraction between partially positive charges and partially negative charges are referred to as **hydrogen bonds**. Water molecules therefore tend to 'stick' together and, as a result, form a lattice of molecules held together by hydrogen bonds. The overall effects of very large numbers of hydrogen bonds are responsible for many of the properties of water.

- Water molecules tend to cling to each other and to other polar molecules. Attraction between water molecules is referred to as **cohesion**; attraction between water molecules and other polar molecules is referred to as **adhesion**.
- Water **stores heat**. A large input of heat energy is needed to raise the temperature of water, which is said to have a high specific heat. As a result, water heats up and cools down relatively slowly.
- A relatively large input of heat energy is needed to convert liquid water into water vapour. Water is therefore said to have a high **latent heat of vaporisation**. The evaporation of 1 g of water requires about 2.43 kJ, so the evaporation of water has a significant cooling effect. This is why the evaporation of sweat is an important mechanism in the regulation of body temperature.
- Water acts as a **solvent**. When sodium chloride (NaCl) dissolves in water, the individual sodium ions (Na^+) and chloride ions (Cl^-) form hydrogen bonds with water molecules. Each ion becomes surrounded by a cloud of water molecules, referred to as a **hydration shell**. Other ions which are important in living organisms, including potassium (K^+), calcium (Ca^{2+}) and magnesium (Mg^{2+}), are also surrounded by a hydration shell. Water molecules form a hydration shell around any molecule which has an electrical charge, whether this is a full electrical charge, as in an ion, or a partial electrical charge, as in a polar molecule. As an example, glucose is a polar molecule because it has slightly polar hydroxyl (-OH) groups. Identify the hydroxyl groups in the glucose molecule shown in Figure A.4. Glucose dissolves in water because water molecules form hydrogen bonds with the polar hydroxyl groups, resulting in the glucose molecules being surrounded by a hydration shell. Hydration shells form around all polar molecules when they dissolve in water, in other words, they are **soluble** in water. Nonpolar molecules, such as fats and oils, are not soluble in water.
- Water **ionises**. The covalent bonds between oxygen and hydrogen in water molecules sometimes break, resulting in the formation of a **proton** (H^+) and a **hydroxide** ion (OH^-). This process is termed **ionisation**. We can represent this process by the chemical equation below.

$$H_2O \rightarrow H^+ + OH^-$$

DEFINITION

Solubility is defined as the maximum concentration that can be obtained of a substance when it is dissolved in a solvent. A **solvent** is a liquid, such as water, in which other substances can dissolve. The substance which dissolves in the solvent is referred to as the **solute**. For example, glucose (the solute) dissolves in water (the solvent) to form a **solution**. Solubility is measured either in grams of solute per 100 g of solvent, or moles of solute per dm^3.

DEFINITION

Many of the important properties of water result from the molecule having electron-rich and electron-poor regions, giving it partially positive and partially negative poles. There are weak forces of attraction between partially positive and negative charges, referred to as **hydrogen bonds**.

> **DEFINITION**
>
> A **mole** is defined as the mass in grams that corresponds to the total atomic mass of all the atoms in a molecule, that is, the molecular mass expressed in grams. For example, water has a molecular mass of 18 (because the atomic mass of hydrogen = 1 and the atomic mass of oxygen = 16), so a mole of water is equivalent to 18 g.

Acidity, the pH scale and buffers

An **acid** is a substance which dissolves in water and dissociates to form H^+ ions. The stronger an acid is, the more it dissociates and the more H^+ ions it produces. The concentration of H^+ ions in solution is used as an indication of acid strength, referred to as the **pH scale**. This scale is based on the slight degree of ionisation of water molecules. In 1 dm^3 (1 litre) of water, about 1 molecule of water in 550 million will be ionised, which corresponds to 1/10 000 000 of a **mole** of H^+ ions.

The molecular mass of H^+ is 1, so a mole of H^+ would weigh 1 g. The molar concentration of H^+ in pure water is more conveniently expressed using a logarithmic scale, by counting the number of zeros after the 1 in the denominator. We write

$$[H^+] = 1/10\ 000\ 000$$

where the square brackets are used to indicate the concentration of H^+ ions. As there are seven zeros after the 1, the molar concentration is 10^{-7} moles per dm^3. The pH is normally expressed as a positive value, so because the molar concentration of H^+ in pure water is 10^{-7}, the pH value is 7. Table A.2 shows some examples of familiar substances, with their pH values and concentrations of hydrogen ions.

> **DEFINITION**
>
> The term **pH** refers to the relative concentration of H^+ ions in solution. The pH is the negative of the logarithm of the molar concentration of H^+ ions. Low pH values (less than 7) indicate high concentrations of H^+ ions (acids) and high pH values (more than 7) indicate low concentrations of H^+ (bases).

Table A.2 *Some examples of familiar substances, with their pH values and concentrations of hydrogen ions*

H^+ ion concentration (moles per dm^3)	pH value	Example
10^{-1}	1 Increasingly acidic	Hydrochloric acid in the stomach
10^{-2}	2	Lemon juice
10^{-3}	3	Vinegar
10^{-7}	7 NEUTRAL	Pure water
10^{-8}	8	Sea water
10^{-11}	11	Household ammonia
10^{-14}	14 Increasingly basic	Sodium hydroxide solution

A **buffer** is a solution which contains a weak acid and a salt of a weak acid, such as ethanoic (acetic) acid and sodium ethanoate (the sodium salt of ethanoic acid). Depending on the relative concentrations of the weak acid and the salt, the solution will have a characteristic pH value.

What is important about buffer solutions is that they resist changes in pH. Thus, for example, if more acid is added to a buffer solution (which will tend to increase the concentration of hydrogen ions and therefore decrease the pH), the excess hydrogen ions are effectively neutralised by the salt of the weak acid, so that the pH changes little, if at all. Similarly, if a base is added to the buffer, this will be neutralised by the weak acid, which, in effect, acts as a store of hydrogen ions.

Buffer solutions are important in living organisms, for example, in maintaining the pH of blood at a value of 7.4. Buffers are also used in enzyme experiments, where it is important to maintain a particular pH value. Changes in pH affect enzyme activity because pH affects the electrical charges in chemical groups within the enzyme molecule. Interactions between these charges affect the shape of the molecule and can affect the efficiency with which the substrate molecule attaches to the active site of an enzyme. An enzyme molecule may be denatured at extreme pH values.

Formulae in chemical equations, oxidation and reduction

A chemical equation is a way of representing a chemical reaction using symbols instead of words to represent the atoms and molecules. We have already used some simple equations, such as that showing the ionisation of a water molecule to form hydrogen ions (H^+) and hydroxide ions (OH^-), which is:

$$H_2O \rightarrow H^+ + OH^-$$

By convention, we write the **reactants** on the left of the equation and the **products** on the right. It is important that the numbers of each atom are the same on both sides of the equation. A more complicated, but familiar, example shows the complete oxidation of glucose ($C_6H_{12}O_6$) in aerobic respiration to form the products carbon dioxide (CO_2) and water (H_2O).

$$C_6H_{12}O_6 + 6O_2 \rightarrow 6CO_2 + 6H_2O$$

What this equation tells us is that complete oxidation of each glucose molecule requires six molecules of oxygen and produces six molecules each of carbon dioxide and water. If you count up the numbers of carbon atoms, hydrogen atoms and oxygen atoms on each side of the equation you will find that they are the same, indicating that the equation is balanced.

Now try balancing the following equation, which shows the overall breakdown of glucose in anaerobic conditions in a yeast cell, producing ethanol (C_2H_5OH) and carbon dioxide (CO_2).

$$C_6H_{12}O_6 \rightarrow C_2H_5OH + CO_2$$

Oxidation is the loss of an electron by an atom, ion or molecule, and is often accompanied by the loss of a hydrogen ion (H^+). The electron lost by one molecule is transferred to another, a process called **reduction**. As an example, respiration involves individual reactions in which the substrate is *oxidised* by the transfer of hydrogen from the substrate to a hydrogen carrier. In this way, the substrate is oxidised and the hydrogen carrier is simultaneously *reduced* when it accepts the hydrogen. We will illustrate this process by reference to a specific reaction which occurs during the Krebs cycle.

$$\text{succinate} + \text{FAD} \rightarrow \text{fumarate} + \text{FADH}_2$$

> **DEFINITION**
> **Oxidation** is the *loss* of an electron or hydrogen by an atom, ion or molecule.
> **Reduction** is the *gain* of an electron or hydrogen.

In this reaction, which is catalysed by the enzyme succinate dehydrogenase, succinate is oxidised by the transfer of two hydrogen atoms from succinate to FAD (flavin–adenine dinucleotide), which is a hydrogen carrier. In this way, succinate is *oxidised* to form fumarate, and the hydrogen carrier, FAD, is simultaneously *reduced* to form $FADH_2$.

Condensation and hydrolysis reactions

A **condensation reaction** occurs when two substances are combined by the removal of water. A covalent bond is formed between the two substances. Condensation reactions occur, for example, when two monosaccharides join to form a disaccharide: for example, glucose and fructose join together by means of a condensation reaction to form sucrose. Such reactions are important because many simple molecules can be joined in this way to form a large molecule, for example a polysaccharide, such as starch. We can illustrate the principle of a condensation reaction by showing in general terms what happens when two substances, represented by A and B, join together.

$$\text{A-OH} + \text{B-H} \xrightarrow{\text{condensation}} \text{A-B} + H_2O$$

Notice that OH is removed from A and H from B; H and OH join together to form water (H_2O). Substances A and B are now joined together, by means of a covalent bond, to form A-B. Figure A.6 shows an example of an actual condensation reaction, when two amino acids join to form a dipeptide.

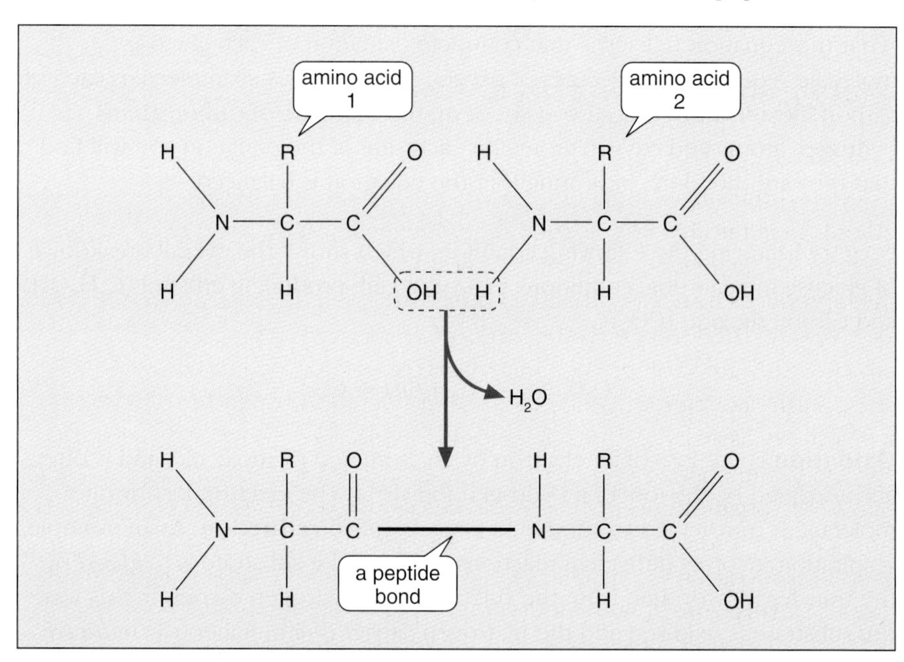

Figure A.6 Two amino acids joining by means of a condensation reaction, forming a dipeptide. The covalent bond between the amino acids is known as a ***peptide bond****. This process can be repeated, progressively adding further amino acids, to build up a polypeptide. Each time a peptide bond is formed, a molecule of water is released.*

Hydrolysis is essentially the reverse of condensation: water molecules are added to break a large molecule into its component sub-units. As examples, hydrolysis of sucrose (a disaccharide) breaks it down into its two component monosaccharides: glucose and fructose. Hydrolysis of a protein would form individual amino acids and hydrolysis of a simple fat would result in the formation of glycerol and fatty acids. Hydrolysis reactions occur during the digestion of carbohydrates, fats and proteins, the reactions are catalysed by enzymes known collectively as **hydrolases**. Figure A.7 summarises the processes of condensation and hydrolysis.

> **DEFINITION**
>
> **Hydrolysis** reactions add water to break large molecules into their smaller sub-units.
> **Condensation** reactions occur when small molecules join together to form large molecules by the removal of water.

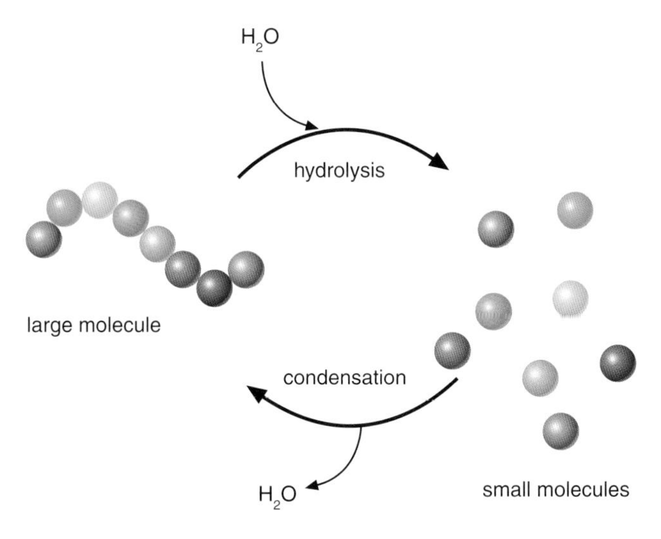

Figure A.7 A large molecule may be broken down into smaller molecules by hydrolyis reactions; small molecules may be joined together by condensation reactions.

> **DEFINITION**
>
> The term **metabolism** is used to describe *all* of the chemical reactions which occur in living cells. Some of these reactions involve the breakdown of complex molecules into simpler ones, with the release of energy. This overall process of breakdown is referred to as **catabolism**.
> **Anabolism** is the term used to describe chemical reactions which join simple molecules to form more complex molecules, such as the condensation reactions joining many monosaccharides to form a polysaccharide.

Energy and chemical reactions

During chemical reactions, energy changes result from the changes in the structure of the substances in the reaction, which involve breaking and reforming chemical bonds. These energy changes take the form of heat. Some reactions produce heat, others absorb heat. A reaction such as combustion, which produces heat, is known as **exothermic**, but a reaction in which heat is taken in is referred to as **endothermic**. Most reactions which occur spontaneously are exothermic. In living organisms, energy produced in a reaction can take various forms, such as heat, light or chemical energy. Heat and light are dissipated, that is, lost from the organism, and therefore are not available to the organism for other purposes. Other forms of energy can be used by the organism to drive other processes. These forms of energy are known as **free energy** and changes in free energy are given the symbol G. A reaction will occur spontaneously if G is negative. In this case, the reaction is termed **exergonic**. If G is positive, an input of free energy will be needed to drive the reaction, which is said to be **endergonic**. In living cells, endergonic reactions include the synthesis of macromolecules, such as proteins, and these reactions are linked to exergonic reactions, which provide the free energy required.

In living cells, most reactions require an input of energy before molecules will react together. This is referred to as the **activation energy**. Imagine a

DEFINITION

There are many different forms of energy, including heat, electrical, light and sound. The SI unit of energy is the **joule** (J), which is defined as the work done when a force of 1 newton (1 N) acts through 1 metre in the direction of the force, that is,
1 joule = 1 newton × 1 metre.
1 joule is approximately the amount of kinetic energy acquired by an apple by the time it hits the ground, after falling off a table. This is a relatively small amount of energy, so we often use the **kilojoule** (kJ), which equals 10^3 joules.

boulder near the top of a steep hill. Before the boulder can roll down the other side of the hill, you have to push it to the top, or in other words put some energy in. This can be likened to the progress of a chemical reaction: the boulder near the top represents the reactants and the energy you have to put in to push the boulder to the top represents the **free energy of activation**. Once the reactants reach this point, the reaction will proceed spontaneously, as the boulder rolls down the hill, and the products are formed. **Enzymes** increase the rates of reactions by reducing the free energy of activation, so that the barrier to a reaction occurring is lower when an enzyme is present.

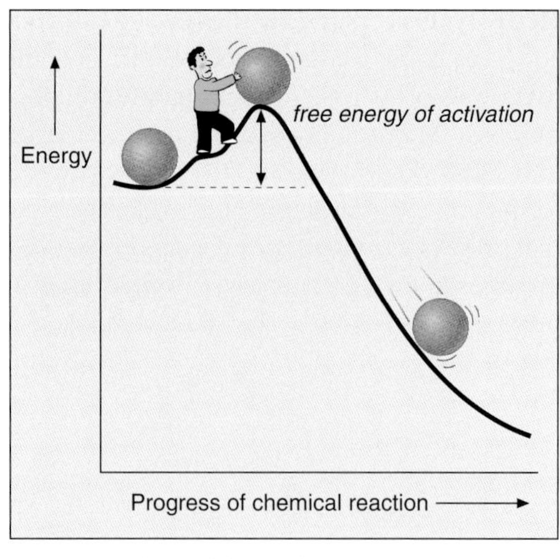

Figure A.8 The boulder analogy of a chemical reaction; activation energy needs to be added to start the reaction.

The electromagnetic spectrum

We are all familiar with visible light. This is a form of *electromagnetic radiation* and represents a small part of a continuous spectrum known as the *electromagnetic spectrum*. This spectrum ranges from very short wavelength γ (gamma) rays, which have wavelengths of about 10^{-12} m, through to very long wavelength radio waves, with wavelengths of about 10^3 m. Visible light is only a small part of the electromagnetic spectrum, and has wavelengths between 4×10^{-7} m (400 nm) and 7×10^{-7} m (700 nm). We see light of different wavelengths as different colours; for example, the colours of a rainbow are seen when white light is split into its individual wavelengths. Red light has a wavelength of about 7×10^{-7} m and merges into the infrared part of the electromagnetic spectrum; violet light has a wavelength of about 4×10^{-7} m and lies next to the ultraviolet part of the spectrum. Figure A.9 represents the electromagnetic spectrum. Although different parts of the spectrum are given names, the boundaries overlap and are not sharply defined.

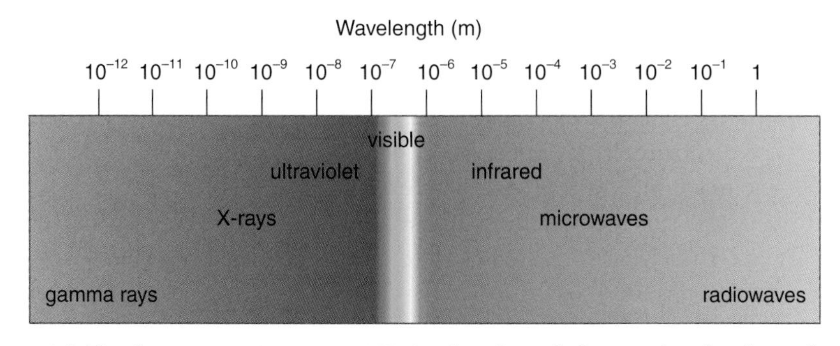

Figure A.9 The electromagnetic spectrum. Notice that the scale for wavelength is logarithmic, that is, equal intervals on the scale represent changes of a factor of ten.

Pressure, partial pressure and diffusion

Dry air consists of 78.09 per cent nitrogen, 20.95 per cent oxygen, 0.93 per cent argon and other inert gases, and 0.03 per cent carbon dioxide. Air exerts a **pressure**, acting downwards, because of the effect of gravity. Pressure is defined as the force per unit area and the SI unit of pressure is the pascal (symbol Pa), which represents a force of 1 newton per m^2. At sea level, air pressure is approximately 100 kilopascals (100 kPa), but air pressure steadily decreases as altitude increases. As an example, at the summit of Mount Everest, atmospheric pressure is reduced to 31.9 kPa. Each individual gas in air exerts a pressure, so the total air pressure represents the total of the **partial pressure** of each individual gas. Therefore, at sea level, the total air pressure of 100 kPa is made up of:

100 kPa × 79.02 per cent = 79.02 kPa of nitrogen

100 kPa × 20.95 per cent = 20.95 kPa of oxygen

100 kPa × 0.93 per cent = 0.93 kPa of argon and other inert gases

100 kPa × 0.03 per cent = 0.03 kPa of carbon dioxide.

The partial pressure of oxygen (written pO_2) is important because it represents the amount of oxygen which is available to a living organism. Oxygen molecules can freely **diffuse** across the cell surface membrane, but do so in solution in water. How many molecules of oxygen can enter a cell depends on the concentration of dissolved oxygen in water. The concentration of a gas dissolved in water depends on four factors:

- the partial pressure of the gas in the air
- the solubility of the gas in water
- the temperature: higher temperatures *decrease* the solubility of gases in water
- solute concentrations: salts reduce the solubility of gas; the solubility of oxygen in seawater at 15 °C is only 85 per cent of its solubility in freshwater.

The uptake of oxygen by a cell depends on the diffusion of oxygen into water. Diffusion is a passive process, dependent entirely on the difference in oxygen concentrations between the interior and exterior of the cell. The rate of diffusion is described by a mathematical relationship known as **Fick's Law of Diffusion**:

$$R = D \times A \times \Delta p / d$$

where:

R = the rate of diffusion
D = the diffusion constant, which depends on the substance through which diffusion is occurring
A = the area across which diffusion takes place
Δp = the difference in partial pressures, or the concentration difference
d = the distance a molecule needs to travel from high concentration to low concentration.

DEFINITION

Pressure is defined as the force per unit area; the SI unit of pressure is the pascal (Pa). The **partial pressure** of a gas is the pressure exerted by one gas in a mixture of gases. For example, at the summit of Mount Everest, the partial pressure of oxygen is 20.09 per cent of 31.9 kPa = 6.4 kPa.

Assessment questions

The following questions have been chosen from recent Unit tests on the content of the Edexcel Biology and Biology (Human) Advanced Subsidiary GCE specification. The style and format of these questions is similar to those that will be set in future tests. The shorter structured questions are designed to test mainly knowledge and understanding of the topics, and the longer questions contain sections in which you may be required to demonstrate skills of interpretation and the evaluation of data. Some sections of the longer questions may require extended answers, for 4 or more marks. From June 2003 there will no longer be a requirement for a 'free prose' answer worth 10 marks.

As many of the topics in this Unit are interlinked, you may find that some questions require knowledge of more than one section of the specification.

Chapter 1 Molecules

1 The table below refers to some disaccharides, their constituent monomers and their roles in living organisms.

Complete the table by writing in the appropriate word or words in the empty boxes.

Disaccharide	Constituent monomers	**One** role in living organisms
Lactose		Carbohydrate source in mammalian milk
	Glucose + glucose	
		Form in which sugars are transported in plants

(Total 5 marks)

(Edexcel GCE Biology (6101/01), January 2001)

2 The diagram below shows the structure of a molecule found in the cell surface membrane.

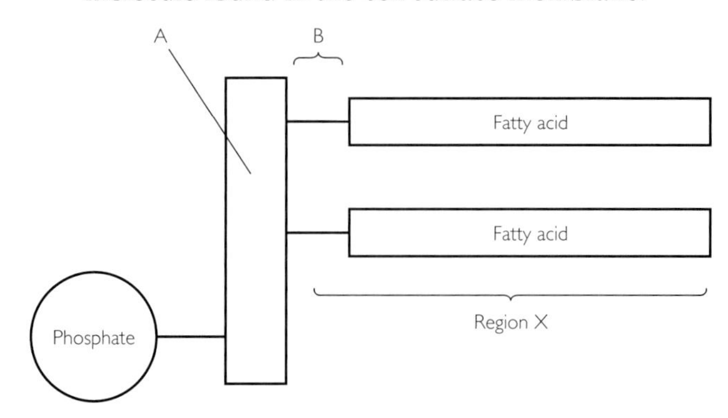

(*a*) Name the type of molecule shown in the diagram. **[1]**

(*b*) Name A and B as labelled on the diagram. **[2]**

(*c*) Region X is said to be **hydrophobic**. What is meant by the term hydrophobic? **[1]**

(*d*) Explain why the cell surface membrane is described as a fluid-mosaic. **[2]**

(Total 6 marks)

(Edexcel GCE Biology (6101/01), June 2001)

3 A solution thought to contain either a reducing sugar or a non-reducing sugar was tested with Benedict's reagent.

(*a*) Describe how the presence of a reducing sugar is detected using Benedict's reagent. **[2]**

(*b*) If the test was negative for reducing sugars, describe what steps you would need to carry out before you could show that a non-reducing sugar was present. **[3]**

(*c*) Describe how Benedict's reagent could be used to compare the concentrations of reducing sugar present in two solutions. **[3]**

(Total 8 marks)

(Edexcel GCE Biology (6101/01), June 2001)

4 Read through the following account of the properties of water, then write on the dotted lines the most appropriate words or words to complete the account.

Water has the chemical formula .... Water molecules are described as .... because they have a slight positive charge at one end of the molecule and a slight negative charge at the other end. As a result, individual molecules form .... bonds with each other.

Water is an important .... in living organisms because most biochemical reactions take place in aqueous solution. Water also has a high ...., which means that its temperature remains relatively stable despite large changes in the temperature of the surrounding environment.

(Total 5 marks)

(Edexcel GCE Biology (6101/01), May 2001)

Chapter 2 Nucleic acids, the genetic code and protein synthesis

1 The diagram below shows part of a molecule of deoxyribonucleic acid (DNA).

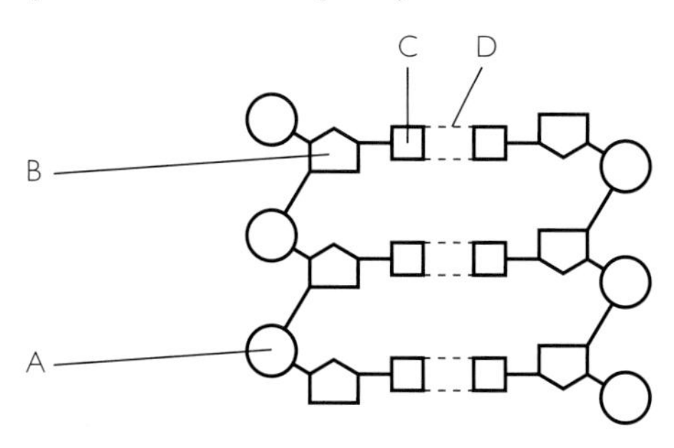

(*a*) Name A, B, C and D. **[4]**

(*b*) Analysis of a molecule of DNA showed that cytosine accounted for 42 per cent of the content of the nitrogenous bases. Calculate the percentage of bases in the molecule which would be thymine. Show your working **[3]**

(*c*) During the process of **transcription**, one of the DNA strands is used as a template for the formation of a complementary strand of messenger RNA (mRNA). The diagram below shows the sequence of bases in part of a strand of DNA.

DNA G C G T C A T G C

mRNA □ □ □ □ □ □ □ □ □

(i) Write the letters of the complementary bases in the boxes of the mRNA strand. **[2]**

(ii) How many amino acids are coded for by this part of the strand of mRNA? **[1]**

(Total 10 marks)

(Edexcel GCE Biology (6101/01) January 2001)

2 Describe the role of messenger RNA (mRNA) in the following processes.

(*a*) Transcription **[3]**

(*b*) Translation **[3]**

(Total 6 marks)

(Edexcel GCE Biology (6101/01), January 2002)

3 The table below refers to the structure of different types of nucleic acids. If the feature is present, place a (✓) in the appropriate box and if the feature is absent place a cross (✗) in the appropriate box.

Feature	DNA	mRNA
Cytosine present		
Uracil present		
Pentose sugar present		
Is single stranded		

(Total 4 marks)

(Edexcel GCE Biology (6101/01), January 2003)

4 The diagram below summarises the steps involved in the semi-conservative replication of DNA.

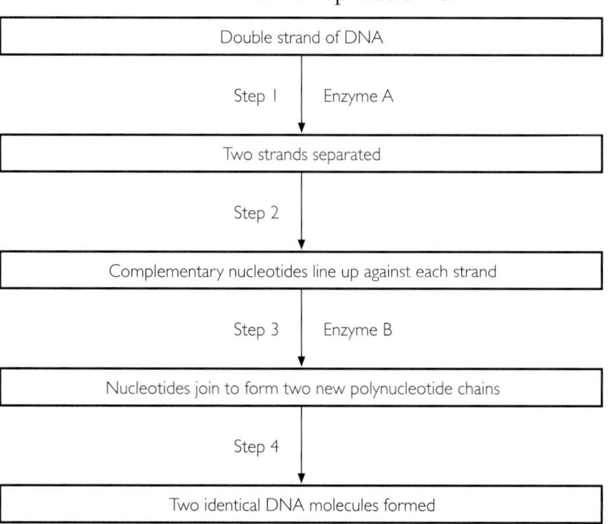

(*a*) Describe how Enzyme A separates the two DNA strands in Step 1. **[1]**

(*b*) In Step 3 the individual nucleotides are joined up to form a polynucleotide chain by Enzyme B.

Name the type of reaction that Enzyme B catalyses. **[1]**

(*c*) Give the phase of the cell cycle during which DNA replication occurs. **[1]**

(*d*) Draw and label a diagram to show the appearance of a chromosome as it appears in metaphase of mitosis. **[3]**

(Total 6 marks)

(Edexcel GCE Biology (6101/01), January 2003)

Chapter 3 Enzymes

1 Sucrase is an enzyme that catalyses the hydrolysis of sucrose. An investigation was carried out to compare the activity of sucrase in solution with immobilised sucrase, over a range of temperatures.

The enzyme in solution was incubated with a solution of sucrose for 5 minutes at different temperatures. The mass of monosaccharide produced at each temperature was determined. This was repeated using immobilised sucrase.

The results of the investigation are shown in the graph.

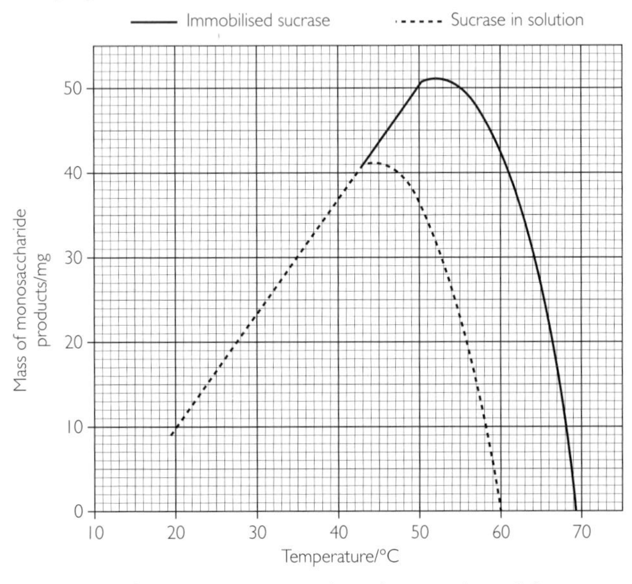

(*a*) Name the **two** monosaccharides produced from the hydrolysis of sucrose. **[2]**

(*b*) What evidence from the graph suggests that the concentrations of sucrase in solution and immobilised sucrase were equivalent? **[1]**

(*c*) Compare the effect of temperature on the activity of sucrase in solution with that on immobilised sucrase. **[3]**

(*d*) Suggest why temperatures above 45 °C have different effects on immobilised sucrase and sucrase in solution. **[2]**

(*e*) Describe how this investigation could be adapted to compare the activity of sucrase in solution with that of immobilised sucrase over a range of pH values. **[4]**

(Total 12 marks)

(Edexcel GCE Biology (6101/01), June 2001)

2 An experiment was performed to determine the effect of pectinase on the yield of apple juice. An apple was cut into small pieces and blended in a food processor to produce apple pulp. The pulp was then left to stand for 15 minutes.

One 50 g sample of pulp was mixed with 5 cm^3 of pectinase solution and a second 50 g sample of pulp was mixed with 5 cm^3 of water. Each sample was then placed in a separate filter funnel and the juice was collected in a measuring cylinder. The volume of juice produced was recorded every minute for 20 minutes.

The results of this experiment are shown in the graph below.

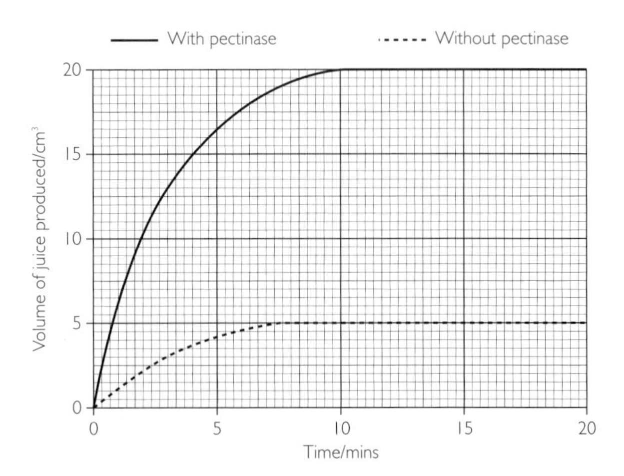

(*a*) (i) With reference to the graph, describe how pectinase affects the production of apple juice. **[2]**

(ii) Explain why pectinase has this effect on the production of apple juice. **[2]**

(*b*) Apples contain chemicals which act as active site-directed inhibitors of pectinase. However, these chemicals lose their effectiveness when exposed to air.

(i) Explain what is meant by the term **active site-directed inhibition**. **[2]**

(ii) In a second experiment the apple pulp was left to stand for 30 minutes before mixing it with pectinase. Suggest what effect this would have had on the volume of juice produced. Explain your answer. **[2]**

(*c*) Describe how you would carry out an experiment to investigate the effect of temperature on the production of apple juice using pectinase. **[4]**

(Total 12 marks)

(Edexcel GCE Biology (6101/01), May 2002)

3 The diagram below represents some of the stages involved in the commercial production of the enzyme protease for use in the manufacture of a biological detergent.

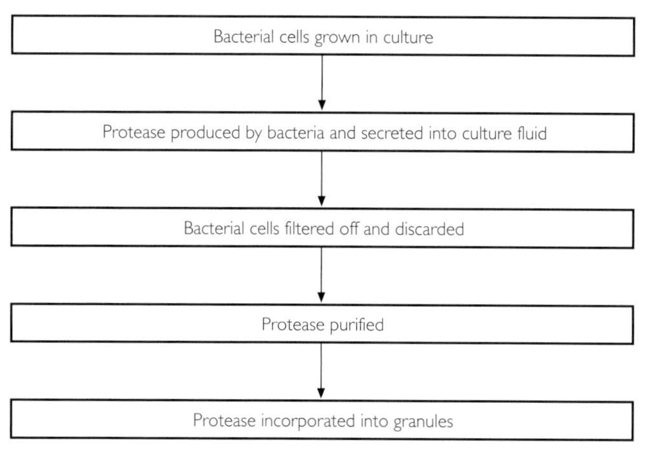

(*a*) Explain why proteases are incorporated into biological detergents. **[3]**

(*b*) The bacteria used in this process normally live in hot water springs where the temperature stays above 45 °C.

Suggest **two** reasons why it is an advantage to use the enzymes from these bacteria in the detergents. **[2]**

(*c*) When the granules containing the protease come into contact with water they swell and release their contents.

Suggest why the enzymes are incorporated into granules during the manufacture of biological detergents. **[2]**

(Total 7 marks)

(Edexcel GCE Biology (6101/01), January 2003)

4 The rate of an enzyme-catalysed reaction can be altered by the presence of an inhibitor.

An investigation was carried out into the effect of an inhibitor on enzyme activity in barley root tips. Enzyme activity was measured by finding the rate at which oxygen was used by the root tips.

Several groups of students carried out experiments in which the volume of oxygen used by root tips of barley seedlings was measured over a period of 2 hours.

Each group of students used 50 root tips from seedlings of the same age. These were placed in the same volume of a pH 6.5 buffer solution and kept at 30 °C.

After 2 hours, the same volume of a 1% solution of an inhibitor was added to each set of root tips. The students continued to record the volume of oxygen used for a further hour.

The results are shown in the graph below.

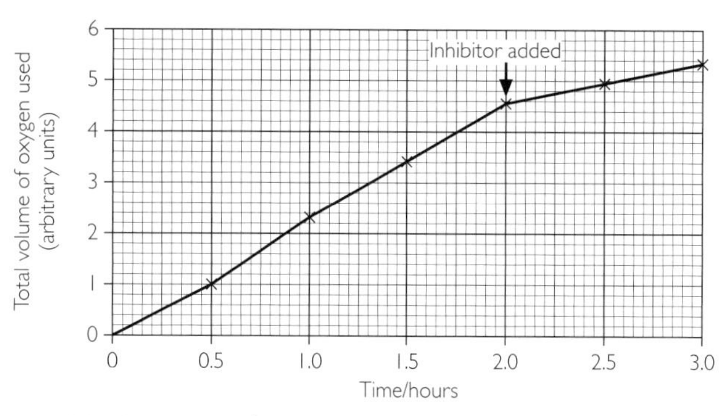

(*a*) Compare the enzyme activity during the first two hours of the experiment with the activity after the inhibitor was added. **[2]**

(*b*) The inhibitor used in this experiment was a non-active site-directed inhibitor.

Explain what is meant by **non-active site-directed inhibition**. **[3]**

(*c*) Suggest what effect increasing the concentration of the inhibitor might have on the rate at which oxygen was used by the root tips. Give an explanation for your answer. **[2]**

(*d*) Suggest why reading were taken for 2 hours before the inhibitor was added. **[1]**

(*e*) Suggest why the root tips were kept at a temperature of 30 °C throughout this investigation. **[2]**

(*f*) Explain why a buffer solution was used in this experiment. **[2]**

(*g*) Suggest **one** reason why the results from the different groups of students might vary. **[1]**

(Total 13 marks)

(Edexcel GCE Biology (6101/01), January 2003)

Chapter 4 Cellular organisation

1 The table below refers to features of prokaryotic and eukaryotic cells. If the feature is usually present, place a tick (✓) in the appropriate box and if the feature is absent, place a cross (✗) in the appropriate box.

Feature	Prokaryotic cell	Eukaryotic cell
Cell surface membrane		
Plasmids		
Ribosomes		
Mitochondria		

(Total 4 marks)

(Edexcel GCE Biology (6101/01), January 2002)

2 A procedure was carried out to separate the major organelles within liver cells. This involved breaking up (homogenising) liver tissue in an ice-cold salt solution which had the same water potential as the cell cytoplasm.

Ultracentrifugation was then used to separate the organelles. Ultracentrifugation is a process that separates materials of different densities by spinning them in a tube at different speeds. The denser materials are forced to the bottom of the tube as a pellet, while less dense materials remain nearer to the top of the tube in liquid known as the supernatant.

The flow chart below summarises the steps involved in this procedure.

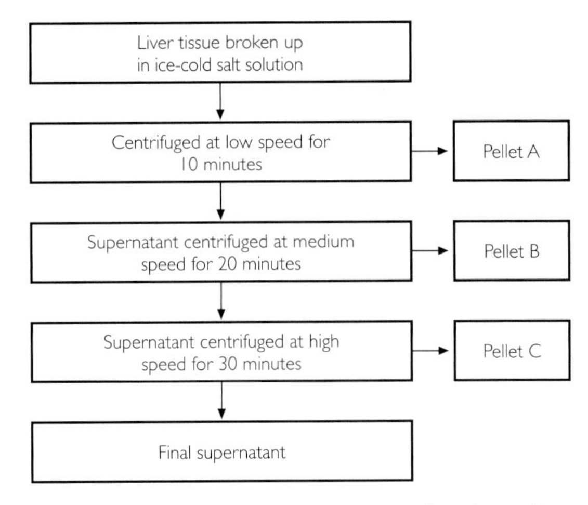

(*a*) Suggest why it was necessary for the salt solution to have the same water potential as the cell cytoplasm. **[2]**

(*b*) This procedure separated mitochondria, nuclei and ribosomes into the three pellets, **A**, **B** and **C**. Complete the table below to show which one of these organelles would be found in which pellet.

Pellet	Organelle
A	
B	
C	

[2]

(*c*) Suggest **two** components of the cell, other than water, that might be present in the final supernatant. **[2]**

(*d*) Draw and label a diagram to show the structure of a mitochondrion. **[4]**

(*e*) Explain why large numbers of mitochrondria are found in liver cells. **[2]**

(Total 12 marks)

(Edexcel GCE Biology (6101/01), January 2002)

3 (*a*) (i) Explain what is meant by the term **facilitated diffusion**. **[2]**

(ii) State **two** ways in which active transport differs from facilitated diffusion. **[2]**

(*b*) In an investigation into the effects of osmosis on red blood cells, seven samples of red blood cells were placed in potassium chloride solutions of different concentrations. After two hours, each sample was examined to find the percentage of cells that had swollen and burst (lysed). The results are shown in the graph below.

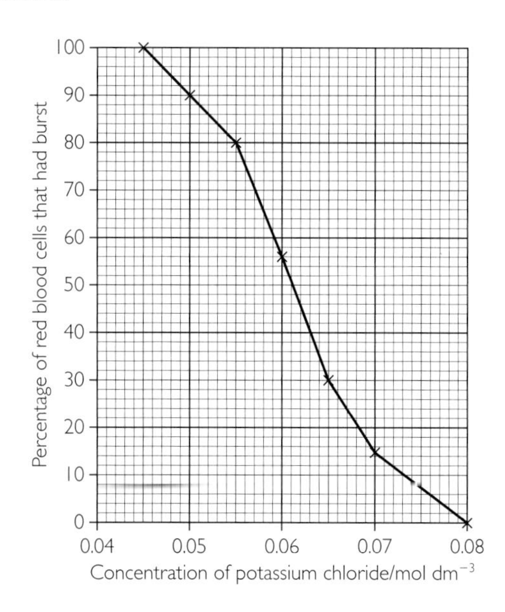

(i) Calculate the difference between the percentage of red blood cells that burst in 0.05 mol dm^{-3} and 0.07 mol dm^{-3} potassium chloride solutions. Show your working. **[2]**

(ii) With reference to water potential, explain why most of the cells burst when placed in 0.05 mol dm^{-3} potassium chloride solution. **[3]**

(iii) Suggest what would happen if red blood cells were placed in a 0.1 mol dm^{-3} solution of potassium chloride. **[1]**

(*c*) Explain why plant cells do not burst when placed in distilled water. **[2]**

(Total 12 marks)

(Edexcel GCE Biology (6101/01), May 2002)

4 The diagram below shows the structure of the cell surface membrane.

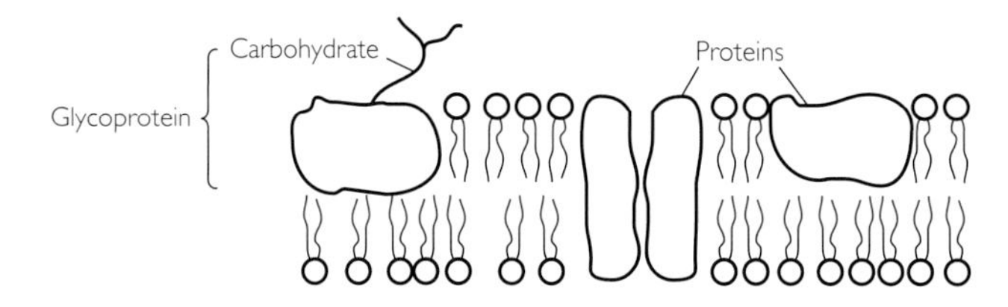

(*a*) The cell surface membrane is composed of a phospholipid bilayer. Explain why the phospholipids in the bilayer are arranged with the fatty acid tails pointing inwards and the phosphate heads outwards. **[3]**

(*b*) The diagram has been magnified three million (3×10^6) times. Calculate the width of the cell surface membrane in μm (micrometres). Show your working. **[3]**

(*c*) State **one** function of each of the following components of the cell surface membrane.

- carbohydrate
- protein **[2]**

(Total 8 marks)

(Edexcel GCE Biology (6101/01), January 2003)

Chapter 5 The cell cycle

1 The diagram below shows the structure of a chromosome as it might appear at the end of prophase of mitosis.

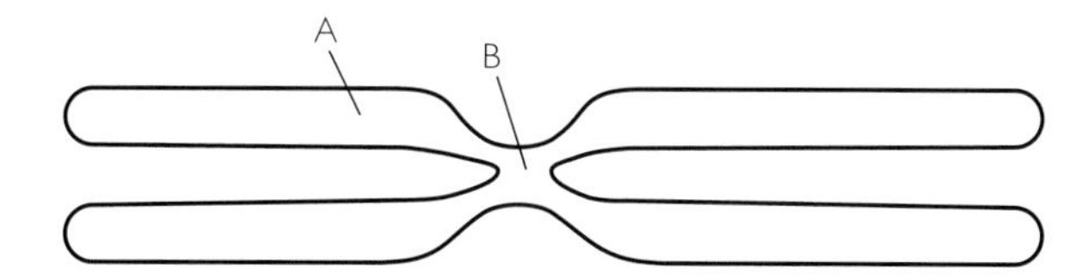

(*a*) Name the parts labelled A and B. **[2]**

(*b*) During metaphase of mitosis, the chromosomes become attached to the equator of the spindle. Name the stage of mitosis that follows metaphase and describe the events that occur in this stage. **[3]**

(*c*) Explain the significance of the stage you have named and described in (b). **[1]**

(*d*) Mitosis forms part of the cell cycle. Name **one** other stage of the cell cycle and state what occurs in the stage that you have named. **[2]**

(Total 8 marks)

(Edexcel GCE Biology (6101/01), January 2002)

ASSESSMENT QUESTIONS

2 (*a*) The table below describes some of the key events that occur during mitosis.

Complete the table by writing the name of the stage of mitosis next to its description.

Key events	Stage
Chromatids separate and move to opposite poles of the dividing cell.	
Chromosomes shorten and thicken. The nuclear envelope breaks down and the spindle forms.	
The spindle fibres break down, the nuclear membrane re-forms and the chromosomes elongate.	
Chromosomes line up on the equator of the cell, attached to spindle fibres by their centromeres.	

[4]

(*b*) The graph below illustrates the change in DNA content during the cell cycle.

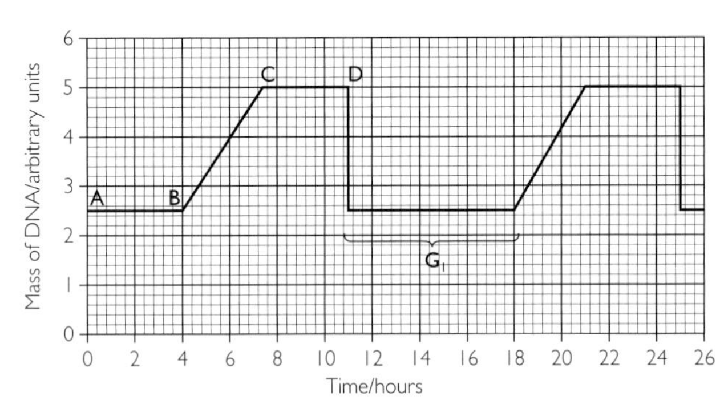

(i) Calculate the percentage of the cell cycle time spent in G_1. **[3]**

(ii) At which point, A, B, C or D, does chromosome replication (the S phase) begin? Explain your answer. **[2]**

(Total 9 marks)

(Edexcel GCE Biology (6101/01), June 2001)

3 The diagram below shows cells from a root tip, prepared by the root tip squash method.

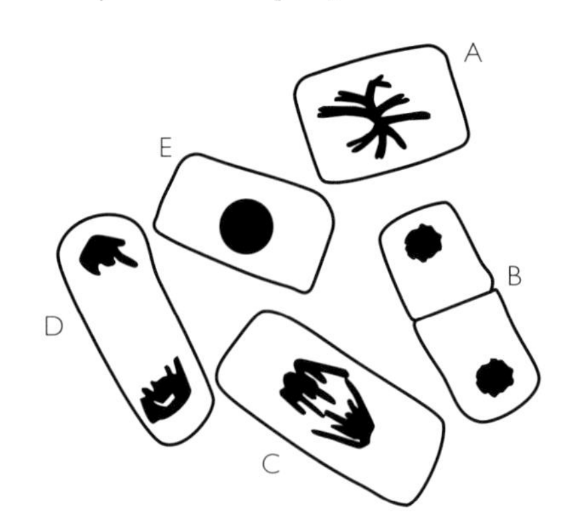

(*a*) Describe how you would prepare a root tip squash so that mitosis can be studied **[4]**

(*b*) State which of the cells labelled A–E is in:

(i) metaphase

(ii) anaphase. **[2]**

(*c*) State **two** events that take place during interphase. **[2]**

(Total 8 marks)

(Edexcel GCE Biology (6101/01), May 2002)

4 Yeast cells grown in culture will divide asexually to form a clone of yeast cells.

The diagram below shows this process occurring.

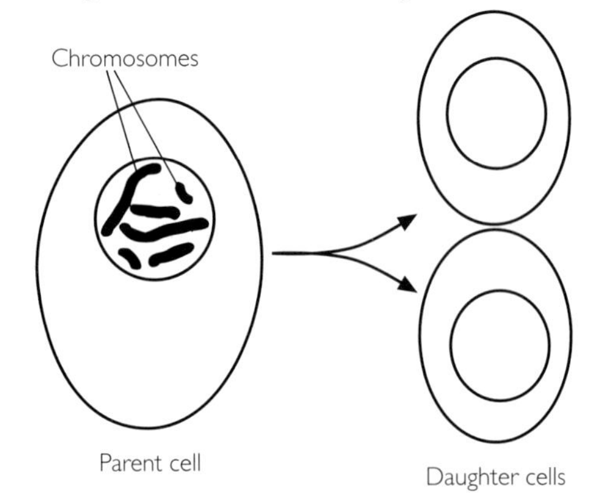

(*a*) Explain what is meant by the term **clone**. **[2]**

(*b*) On the diagram of **one** of the daughter cells, draw in the chromosomes that would be found in the nucleus. **[1]**

(*c*) Suggest two advantages of asexual reproduction to an organism. **[2]**

(Total 5 marks)

(Edexcel GCE Biology (6101/01) January 2003)

Mark Schemes

In the mark schemes, the following symbols are used.
; indicates separate marking points
/ indicates alternative marking points
eq. means correct equivalent points are accepted
{} indicate a list of alternatives

Chapter 1 Molecules

1

Disaccharide	Constitutent monomers	**One** role in living organisms
	Glucose and galactose ;	
Maltose ;		Energy / food source in germinating *seeds;*
Sucrose ;	Glucose and fructose ;	

(Total 5 marks)

2 (*a*) Phospholipids; **[1]**

(*b*) A Glycerol / propan 1,2,3 triol ; **[1]**

B Ester bond / ester linkage ; **[1]**

(*c*) Insoluble in water / does not dissolve in water / non-polar ; **[1]**

(*d*) (Fluid because) phospholipids move (around membrane) ; **[1]**

(Mosaic because) membrane contains proteins / glycoproteins (lying amongst phospholipids / eq ; **[1]**

(Total 6 marks)

3 (*a*) (Solution) heated / boiled (with Benedict's solution / reagent ;

Green / yellow / orange / red / brown (precipitate) ; **[2]**

(*b*) Heat with acid *or* add sucrase ;

Neutralise *or* incubate [if using enzyme] ;

(Heat with) Benedict's / repeat test / eq ; **[3]**

(*c*) Same / stated, volumes of each (test) solution; Same / stated, volume Benedict's solution (to each) ; Stated / same, time / temperature (for heating) / boil in waterbath ; Weigh precipitate / colour comparison / reference to colorimeter comparison / time taken to reach standard or same colour / reference to rate of colour change ; **[3]**

(Total 8 marks)

1 H_2O ; **[1]**

(di)polar ; **[1]**

Hydrogen / H ; **[1]**

Solvent ; **[1]**

(specific) heat capacity ; **[1]**

(Total 5 marks)

Chapter 2 Nucleic acids, the genetic code and protein synthesis

1 (*a*) A Phosphate ;
B Deoxyribose ;
C (organic) base / thymine / adenine ;
D hydrogen bond / H bond ; **[4]**

(*b*) Total percentage of C + G = 84% ;
therefore T will be (100 – 84) ÷ 2 ; = 8% ; **[3]**

(*c*) (i) C G C \ A G U \ A C G ; ; **[2]**

[All correct = 2 marks, 1 error = 1 mark]

(ii) 3 ; **[1]**

(Total 10 marks)

2 (*a*) (mRNA) is a copy of DNA ;
(Copy of) part of DNA / eq ;
(Copy of) one strand / sense strand ;
mRNA is complementary (to DNA) / mRNA made up of complementary bases ;
mRNA strand, built / formed (looking for idea that mRNA strand is put together during the process) / reference to enzyme ;
Carries genetic code to, cytoplasm / out of nucleus / to ribosome ; **[3]**

(*b*) Genetic information / base sequence / code, in mRNA determines amino acid sequence ;
Codons / base triplet on mRNA ;
Determines amino acid ;

(Codons) pair with, *complementary* triplet / anticodons, on tRNA ;
Reference to start / stop codons / sequences / binding sequences ;
Occurs on ribosomes ; **[3]**

(Total 6 marks)

3

Component	DNA	mRNA
Cytosine present	✓	✓
Uracil present	✗	✓
Pentose sugar present	✓	✓
Is single stranded	✗	✓

[Any two correct = 1 mark]

(Total 4 marks)

4 (*a*) Breaks the hydrogen bonds (between the strands) ; **[1]**

(*b*) Condensation / polymerisation ; **[1]**

(*c*) Interphase / S phase / synthesis phase ; **[1]**

(*d*) Correct diagram, as below ;

[Accept constriction for centromere]
[Accept single line for chromatid]

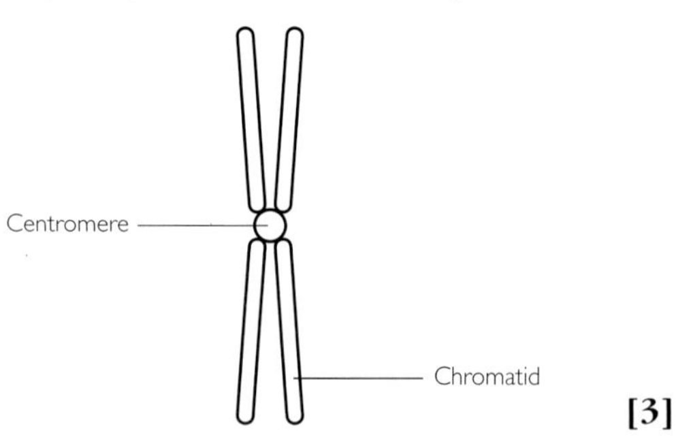

[3]

(Total 6 marks)

1 (*a*) Glucose ;

Fructose ; **[2]**

(*b*) Same slope / mass of products, at lower temperatures / below 43 °C ; **[1]**

(*c*) (Activity of) both increases as temp. increases up to 45 °C ; Credit manipulated figures from graph up to 43 °C ; Peak / optimum, at 43 to 45 °C / lower for solution *and* 51 to 52 °C / higher for immobilised; Comparison of figures between 42 °C and 60 °C ; Activity ceases at 60 °C / lower for solution *and* 69 to 70 °C / higher for immobilised ; **[3]**

(*d*) Immobilised enzyme / more stable (at high temperatures) ;

Because held in position / eq ; **[2]**

(*e*) Constant / stated, temperature / time for both; (below 43 °C) ; Same / stated volume / concentration, of sucrose solution / substrate ; Same / stated volume / concentration, of enzyme / sucrose ; Use *buffers* over range of pH ; Mass of products determined ; Plot graph of mass of products against pH for sucrose solution *and* immobilised sucrase ; **[4]**

(Total 12 marks)

2 (*a*) (i) The production of juice is increased / more juice is produced ; Four times as much / 15 cm^3 more juice was produced ; The rate of production increases / the graph has a steeper line with prectinase ; Production lasts longer ; **[2]**

(ii) Enzyme digests/breaks down/ hydrolyses pectin ; In cell walls ;

Allowing release of juice / reference to permeability ; **[2]**

(*b*) (i) Inhibitor has similar shape to, substrate / pectin ; Binds to active site ; Prevents entry of / competition with, substrate ; **[2]**

(ii) More juice released / juice released faster Increased enzyme activity / decreased inhibition ; **[2]**

(*c*) 1. Use same type / variety of apples ; 2. Same / stated mass / volume of pulp ; 3. Range of temperatures / at least three stated temperatures ; 4. Same / stated volume / concentration of enzyme / pectinase ; 5. Same time / stated time period ; 6. Measure volume of juice extracted ; 7. Plot graph of volume of apple juice vs. temperature ; **[4]**

(Total 12 marks)

3 (*a*) To {remove / digest} {protein / named example / blood / gravy / eggs} ; (Proteins broken down to) peptides / amino acids ; Correct reference to solubility ; Less {heat / energy} required / lower temperature needed / less damage to material ; **[3]**

(*b*) Do no denature at temperatures that the detergents work at / greater stability ; Optimum temperature is higher / works faster than other enzymes at higher temperature ; Stains easier to shift at high temperatures ; Have longer shelf-life / eq ; **[2]**

(*c*) Enzymes will only {be released / start working} {during washing procedure / when needed} ;

Less danger of harm to users ; Increases stability of the enzyme ; **[2]**

(Total 7 marks)

4 (*a*) Less active after inhibitor is added :

Comparative use of figures (calculation of rates) ; **[2]**

(*b*) Inhibitor attaches to part of enzyme molecule {other than active site / allosteric site} ; Alters shape of active site / eq ; So substrate cannot bind / enzyme-substrate complex cannot form / eq ; Slows rate of reaction ; **[3]**

(*c*) Decrease rate further ; Because it will affect *more* of the enzyme molecules ;

OR

Stop reaction ; Because it will affect *all* of the enzyme molecules ; **[2]**

(*d*) So that rate of reaction without the inhibitor was known / {acclimatisation / equilibration idea} / to compare results before and after inhibitor was added ; **[1]**

(*e*) {Suitable / optimum} temperature for the activity of the enzyme ; (fluctuations) would alter rate of oxygen uptake ; (fluctuations) would alter (rate of) enzyme activity ; **[2]**

(*f*) To keep the pH constant ; Fluctuations in pH could affect enzyme activity ; Variations in pH could alter the ionic charges ; Substrate molecules could be prevented from binding at the active site ; **[2]**

(*g*) The concentration of the enzyme in the root tips might not be the same / difficult to make sure that the enzyme concentration is the same in all the samples / size of root tips vary / different stages of development / different metabolic rates / different quantities of stored energy / reference to different genotypes / kept in different conditions before experiment ; **[1]**

(Total 13 marks)

Chapter 4 Cellular organisation

1

Feature	Prokaryotic cell	Eukaryotic cell
Cell surface membrane	✓	✓
Plasmids	✓	✗
Ribosomes	✓	✓
Mitochondria	✗	✓

Any two correct boxes for one mark

(Total 4 marks)

2 (*a*) Prevent, entry / exit, of water ;

By osmosis ;

Which may affect / eq, organelles ; **[2]**

(*b*)

Pellet	Organelle
A	nuclei ;
B	mitochondria ;
C	ribosomes ;

[2]

(*c*) Glucose / monosaccharides / disaccharides / sugar ;

Glycogen ;

Proteins / (poly)peptides ;

Enzymes / suitable named example ;

Ions / named example ;

Amino acids / named example ;

Lipids / phospholipids / triglycerides / fats ;

Microtubules / centrioles ;

Microfilaments ;

RNA / mRNA / tRNA ;

[Accept other correct substances present in liver cells] **[2]**

(*d*) Two membranes shown [**M**] ;

Inner membrane shown folded [**F**] ;

Membrane / envelope ;

Intermembranal space ;

Cristae ;

Matrix ;

Ribosomes / DNA, in matrix ;

Stalked particles / ATPase ; **[4]**

(*e*) To produce (large amounts) of ATP ;

By, *aerobic* respiration / Krebs cycle / electron transport chain ;

(Because) liver cells are (metabolically) very active / eq ; **[2]**

(Total 12 marks)

3 (*a*) (i) Movement down concentration gradient / eq :

Involves protein in membrane ; **[2]**

(ii) (Active transport requires ATP energy ;

(Active transport) occurs against/up the concentration gradient ; **[2]**

(*b*) (i) (0.05 mol dm^{-3} =) 90% and (0.07 mol dm^{-3} =) 15% ;
[allow 14% reading, no consequential error] **[2]**

(ii) Water potential of solution is more than that of red blood cells / allow converse ; Water enters the cells ; By osmosis / down water potential gradient ; Expansion of cytoplasm / eq / increasing pressure / stretching the membrane ; **[3]**

(iii) Shrink / shrivel up / crenate / eq / water would move out / exosmosis ; **[1]**

(*c*) Cell wall ;

Resists expansion / eq ; **[2]**

(Total 12 marks)

4 (*a*) {Fatty acid / tails} are {hydrophobic / non-polar} ; {Phosphate / heads} are {hydrophilic / polar} ; (so can) interact with {water / polar environment} ; **[3]**

(*b*) Correct measurement ; [17 mm or 24 mm] Divide by magnification ;

Correct conversion to μm ; [answer = 0.0057 or 0.008] **[3]**

(*c*) Carbohydrate: *Cell* recognition / cell adhesion / eq ;

Protein: Transport of molecules / eq OR receptor for hormone / eq OR enzymes ; **[2]**

(Total 8 marks)

Chapter 5 The cell cycle

1 (*a*) **A** – chromatid ;

B – centromere **[2]**

(*b*) Anaphase ; Chromatids separate / centromere splits ; Move / OR pulled to (opposite) poles / OR ends of cells / OR ends of spindle / to centrioles ; By spindle fibres / microtubules ; **[3]**

(*c*) Daughter cells *genetically* identical to parent cell / maintains chromosome number / eq ; **[1]**

(*d*) Interphase/G1/S/G2/cytokinesis / cleavage ; **If interphase / named stage** – growth / synthesis of organelles / synthesis / OR replication of DNA / division of organelles ; OR if Cytokinesis – division of the cytoplasm / formation of cell plate in plants ; Points linked **[2]**

(Total 8 marks)

2 (*a*) Anaphase ;

Prophase ;

Telophase ;

Metaphase ; **[4]**

(*b*) (i) (G lasts) 7.0 hrs (and cell cycle time is 14 hrs) / 18–11 ;

$7 \times 10 \div 14$;

= 50% ; **[3]**

(ii) B ;

DNA replication / DNA mass beginning to, increase / double ; **[2]**

(Total 9 marks)

3 (*a*) Reference to named stain (acetic orcein / acetocarmine / Feulgens / Schiffs) Warm / heat ; Break open tip with (mounted) needle / eq ; Mount in stain / acid / water ; (gently) squash under coverslip / eq ; **[4]**

(*b*) (i) A ;

(ii) C ; **[2]**

(*c*) Synthesis / division / multiplication of organelles (or named organelle) ; Growth ; Replication of DNA / chromosomes ; Protein synthesis / name of specific protein being synthesised ; Any normal cell activities, named example (e.g. respiration) ; **[2]**

(Total 8 marks)

4 (*a*) Genetically identical ;

(identical) to {each other / the parent cell} ; **[2]**

(*b*)

[1]

(*c*) Daughter cells {suitable / eq} for surviving in the present conditions / preserves desirable characteristics / eq ; Rapid / eq ; Only one parent is needed ; **[2]**

(Total 5 marks)

Index

Page references in *italics* refer to a table or an illustration.

INDEX

INDEX